FDA Establishment Inspections

Pharmaceutical, Biotechnology, Medical Device and Food Manufacturing

Concise Reference

FDA Establishment Inspections

Pharmaceutical, Biotechnology, Medical Device and Food Manufacturing

Concise Reference

FDA Establishment Inspections: Pharmaceutical, Biotechnology, Medical Device and Food Manufacturing Concise Reference

PharmaLogika, Inc.
PO Box 461
Willow Springs, NC 27592

www.pharmalogika.com

Author / Editor: Mindy J. Allport-Settle

Published by PharmaLogika, Inc.

Printed in the United States of America. First Printing.

ISBN 0-9821476-6-X
ISBN-13 978-0-9821476-6-5

Contents

Preface 1
  About this Book 1
    Included Documents and Features 1
    Reference Tools 1
  About the Reference Tools 2
    Overview and Orientation 2
    Combined Glossary 2
    Combined Index for all chapters and
    addendums 2

Overview and Orientation 3
  The Food and Drug Administration (FDA) 3
    Historical Origins of Federal Food and
    Drug Regulation 3
    The 1906 Food and Drug Act and
    creation of the FDA 4
    The 1938 Food, Drug, and Cosmetic
    Act 5
    Early FD&C Act amendments:
    1938-1958 6
    Good Manufacturing Practices vs.
    Current Good Manufacturing Practices 7
    Organizational Structure 7
    How does the FDA communicate with
    Industry? 8
      Code of Federal Regulations 8
      Guidance Documents 8
      Federal Register 9
      Direct Communication and Letters 9
      www.FDA.gov 9
      Conferences 10
    False Statement to a Federal Agency 10
    CFR Title 21 - Food and Drugs:
    Parts 1 to 1499 10
      General 10
      Foods 12
      Drugs 14
    New Drugs and Over-the-Counter Drug
    Products 15
    Veterinary Products 16
    Biologics 18

Cosmetics 18
Medical Devices 18
Mammography 20
Radiological Health 20
Regulations under Certain Other Acts 20
Controlled Substances 20
Office of National Drug Control Policy 21

Part I

Investigations Operations Manual 23

Foreword 25
Vision / Mission / Values 26
Vision 26
Mission 26
Values 26

Investigations Operations Manual Chapter 5: Establishment Inspections 29

5.1 - Inspection Information 29
5.1.1 - Authority to Enter and Inspect 29
5.1.1.1 - FDA Investigator's
Responsibility 30
5.1.1.2 - Credentials 30
5.1.1.3 - Written Notice 30
5.1.1.4 - Written Observations 30
5.1.1.5 - Receipts 31
5.1.1.6 - Written Demand for
Records 31
5.1.1.7 - Written Requests for
Information 31
5.1.1.8 - Business Premises 32
5.1.1.9 - Premises Used for Living
Quarters 32
5.1.1.10 - Facilities where Electronic
Products are Used or Held 32
5.1.1.11 - Multiple Occupancy
Inspections 33
5.1.1.12 - Authority for Examinations
and Investigations 33
5.1.1.13 - Authority to Implement
Section 702(e)(5) of the FD&C Act 33

5.1.1.14 - Products Imported Under the Provisions of Section 801(d)(3) of the FD&C Act 36
5.1.2 - Inspectional Approach........................ 38
    5.1.2.1 - Depth of Inspection.................. 38
    5.1.2.2 - Inspection Walk Through 39
    5.1.2.3 - Signing Non-FDA Documents 39
    5.1.2.4 - Technical Assistance............... 40
    5.1.2.5 - Team Inspections 40
5.1.3 - Inspection of Foreign Firms............... 43
5.1.4 - Inspectional Precautions 43
    5.1.4.1 - Clothing................................. 44
    5.1.4.2 - PHS Recommendations - Basic Sanitary Practices 45
    5.1.4.3 - Representatives Invited by the Firm to View the Inspection.......... 45
5.1.5 - General Procedures & Techniques 46
    5.1.5.1 - Candling................................. 46
    5.1.5.2 - Label Review 47
    5.1.5.3 - Field Exams 47
5.2 - Inspection Procedures 48
5.2.1 - Pre-Inspectional Activities 48
    5.2.1.1 - Pre-Announcements 49
    5.2.1.2 - Personal Safety 53
    5.2.1.3 - FACTS Personal Safety Alert.. 56
    5.2.1.4 - Personal Safety 57
5.2.2 - Notice of Inspection........................... 60
    5.2.2.1 - Multiple Date Inspections.......... 61
    5.2.2.2 - Inspection of Vehicles.............. 62
    5.2.2.3 - Follow-Up Inspections by Court Order 62
    5.2.2.4 - Conducting Regulatory Inspections When the Agency is Contemplating Taking, or is Taking, Criminal Action 64
    5.2.2.5 - When Evidence of a Criminal Violation is Discovered in the Course of a Regulatory Inspection... 65
    5.2.2.6 - Use of Evidence Gathered in the Course of a Criminal Investigation 66

5.2.2.7 - Use of Evidence Voluntarily
Provided to the Agency 67
5.2.2.8 - Concurrent Administrative,
Civil, and Criminal Actions...................... 67
5.2.2.9 - Working with a Grand Jury 68
5.2.3 - Reports of Observations 68
5.2.3.1 - Preparation Of Form
FDA 483.. 70
5.2.3.2 - Reportable Observations 76
5.2.3.3 - Non-Reportable
Observations 78
5.2.3.4 - Annotation of the FDA 483....... 80
5.2.3.5 - Government Wide Quality
Assurance Program (GWQAP) 82
5.2.3.6 - Distribution of the FDA 483 82
5.2.4 - Receipt - Factory Samples 83
5.2.4.1 - Items Requiring Receipt........... 83
5.2.4.2 - Items Not Requiring
Receipt.. 83
5.2.5 - Inspection Refusal 84
5.2.5.1 - Refusal of Entry....................... 84
5.2.5.2 - Refusal to Permit Access
to or Copying of Records....................... 85
5.2.5.3 - Refusal after Serving
Warrant.. 85
5.2.5.4 - Hostile and Uncooperative
Interviewees 85
5.2.6 - Inspection Warrant............................. 89
5.2.7 - Discussions with Management.......... 90
5.2.7.1 - Protection of Privileged
Information.. 91
5.2.7.2 - Refusals of Requested
Information.. 92
5.2.8 - Consumer Complaints 92
5.2.9 - Interviewing Confidential
Informants 92
5.2.9.1 - How to handle the first
contact 93
5.2.9.2 - Protect the identity of the
source 95
5.2.10 - Routine Biosecurity Procedures
for Visits to Facilities Housing or
Transporting Domestic or Wild Animals........ 97
5.2.10.1 - Pre-Inspection Activities......... 98

5.2.10.2 - General Inspection Procedures 100

5.2.10.3 - Special Situation Precautions 102

5.3 - Evidence Development 103

5.3.1 - Techniques 103

5.3.2 - Factory Samples 103

5.3.3 - Exhibits 104

5.3.4 - Photographs 105

5.3.4.1 - In-Plant Photographs 107

5.3.4.2 - Photo Identification and Submission 107

5.3.4.3 - Preparing and Maintaining Digital Photographs as Regulatory Evidence 112

5.3.4.4 - Preparing Digital Photos for Insertion in a Turbo Establishment Inspection Report (EIR) 114

5.3.4.5 - Photograph Requests 115

5.3.5 - Recordings 116

5.3.6 - Responsible Individuals 117

5.3.6.1 - Discussion on Duty, Power, Responsibility 117

5.3.6.2 - Inspection Techniques How to Document Responsibility 118

5.3.7 - Guarantees and Labeling Agreements 120

5.3.7.1 - Guaranty 121

5.3.7.2 - Labeling Agreement 121

5.3.7.3 - Exemption Requirements 121

5.3.8 - Records Obtained 121

5.3.8.1 - Identification of Records 122

5.3.8.2 - Identifying Original Paper Records 122

5.3.8.3 - Filmed or Electronic Records ... 124

5.3.8.4 - Requesting and Working with Computerized Complaint and Failure Data 127

5.3.8.5 - Listing of Records 132

5.3.8.6 - Patient and/or Consumer Identification on Records 132

5.3.9 - Request for Sample Collection 133

5.3.9.1 - FACTS Assignment Section 134

5.3.9.2 - FACTS Operations Section 134

5.3.9.3 - FACTS Organizations
Section... 135
5.3.10 - Post-Inspection Notification
Letters 135
5.4 - Food 136
   5.4.1 - Food Inspections 136
      5.4.1.1 - Preparation and References 136
      5.4.1.2 - Inspectional Authority 137
      5.4.1.3 - Records Access Under
      BT Authority 139
      5.4.1.4 - Food and Cosmetic
      Defense Inspectional Activities.............. 140
      5.4.1.5 - Food Registration 146
      5.4.1.6 - CFSAN Bio-research
      Monitoring 150
   5.4.2 - Personnel............................... 150
      5.4.2.1 - Management 150
      5.4.2.2 - Employees.............................. 151
   5.4.3 - Plants and Grounds 152
      5.4.3.1 - Plant Construction,
      Design and Maintenance........................ 152
      5.4.3.2 - Waste Disposal 152
      5.4.3.3 - Plant Services 153
   5.4.4 - Raw Materials 153
      5.4.4.1 - Handling Procedure 154
      5.4.4.2 - Condition 154
      5.4.4.3 - Food Chemicals Codex............ 154
   5.4.5 - Equipment and Utensils...................... 155
      5.4.5.1 - Filtering Systems..................... 155
      5.4.5.2 - Sanitation of Machinery........... 155
      5.4.5.3 - Conveyor Belt Conditions........ 155
      5.4.5.4 - Utensils.................................... 156
      5.4.5.5 - Mercury and Glass
      Contamination 156
      5.4.5.6 - UV Lamps............................... 156
      5.4.5.7 - Chlorine Solution Pipes........... 156
      5.4.5.8 - Sanitation Practices................. 156
   5.4.6 - Manufacturing Process 157
      5.4.6.1 - Ingredient Handling 158
      5.4.6.2 - Formulas 158
      5.4.6.3 - Food Additives......................... 158
      5.4.6.4 - Color Additives 160
      5.4.6.5 - Quality Control......................... 161
      5.4.6.6 - Packaging and Labeling........... 162

5.4.7 - Sanitation 163
    5.4.7.1 - Routes of Contamination 164
    5.4.7.2 - Microbiological Concerns......... 167
    5.4.7.3 - Storage 169
5.4.8 - Distribution 171
    5.4.8.1 - Promotion and Advertising........ 171
    5.4.8.2 - Recall Procedure 172
    5.4.8.3 - Complaint Files....................... 172
5.4.9 - Other Government Inspection 172
    5.4.9.1 - Federal................................... 173
    5.4.9.2 - State and Local....................... 173
    5.4.9.3 - Grade A Dairy Plant
    Inspections.. 174
5.4.10 - Food Standards............................. 175
    5.4.10.1 - Food Establishment
    Inspection.. 175
    5.4.10.2 - Food Inspection Report 176
    5.4.10.3 - Violative Inspections.............. 177
5.5 - Drugs 178
5.5.1 - Drug Inspections 178
    5.5.1.1 - Preparation and References..... 179
    5.5.1.2 - Inspectional Approach 181
    5.5.1.3 - CDER Bio-research
    Monitoring 182
5.5.2 - Drug Registration & Listing 182
5.5.3 - Promotion and Advertising 183
5.5.4 - Guarantees and Labeling
Agreements 183
    5.5.5 - Other Inspectional Issues 183
    5.5.5.1 - Intended Use 183
    5.5.5.2 - Drug Approval Status............... 184
    5.5.5.3 - Drug Status Questions............. 184
    5.5.5.4 - Drug/Dietary Supplement
    Status 184
    5.5.5.5 - Approved Drugs....................... 184
    5.5.5.6 - Investigational Drugs 184
    5.5.5.7 - Clinical Investigators
    and/or Clinical Pharmacologists............. 185
5.5.6 - CDER Bio-Research Monitoring 185
5.5.7 - Adverse Event Reporting 185
5.5.8 - Drug Inspection Report 186
5.6 - Devices 186
5.6.1 - Device Inspections 186
    5.6.1.1 - Technical Assistance............... 187

5.6.1.2 - Sample Collection During
Inspection 187
5.6.1.3 - Types of Inspections 188
5.6.1.4 - CDRH Bio-research
Monitoring 188
5.6.2 - Medical Device Quality System/
Good Manufacturing Practices 189
5.6.2.1 - Pre-Inspectional Activities 190
5.6.2.2 - Quality Audit 191
5.6.2.3 - Records 192
5.6.2.4 - Complaint Files 192
5.6.3 - Sterile Devices 194
5.6.4 - Labeling 194
5.6.5 - Government-Wide Quality
Assurance Program (GWQAP) 194
5.6.6 - Contract Facilities 195
5.6.7 - Small Manufacturers 196
5.6.8 - Banned Devices 196
5.6.9 - Device Inspection Reports 197
5.7 - Biologics 197
5.7.1 - Definition 197
5.7.2 - Biologics Inspections 197
5.7.2.1 - Authority 198
5.7.2.2 - Donor Confidentiality 200
5.7.2.3 - Inspectional Objectives 201
5.7.2.4 - Preparation 201
5.7.2.5 - Inspectional Approach 202
5.7.2.6 - Regulations, Guidelines,
Recommendations 203
5.7.2.7 - Technical Assistance 204
5.7.2.8 - CBER Bio-research
Monitoring 204
5.7.3 - Registration, Listing and
Licensing 204
5.7.3.1 - Registration and Listing 204
5.7.3.2 - MOUs 206
5.7.3.3 - Biologic License 207
5.7.3.4 - Approval of Biological
Devices 207
5.7.4 - Responsible Individuals 207
5.7.5 - Testing Laboratories 208
5.7.6 - Brokers 208
5.8 - Pesticides 209
5.8.1 - Pesticide Inspections 209
5.8.2 - Current Practices 210

5.8.3 - Growers..212
    5.8.3.1 - Pesticide Application................212
    5.8.3.2 - Pesticide Misuse/Drift/Soil
    Contamination.......................................213
5.8.4 - Packers and Shippers.......................214
5.8.5 - Pesticide Suppliers..........................215
5.8.6 - Pesticide Applicators.......................215
5.8.7 - Sample Collections..........................215
5.9 - Veterinary Medicine...............................215
5.9.1 - CVM Website..................................215
5.9.2 - Veterinary Drug Activities.................216
5.9.3 - Medicated Feeds and Type A
Articles..217
5.9.4 - BSE Activities.................................217
5.9.5 - Tissue Residues..............................218
5.9.6 - Veterinary Devices..........................219
5.9.7 - Animal Grooming Aids.....................220
5.9.8 - CVM Bio-Research Monitoring...........220
5.10 - Reporting..220
5.10.1 - Establishment Inspection Report
(EIR)..220
5.10.2 - Endorsement.................................221
    5.10.2.1 - Compliance Achievement
    Reporting System (CARS).....................223
5.10.3 - Facts Establishment Inspection
Record (EI Record)....................................224
5.10.4 - Narrative Report............................225
    5.10.4.1 - Non-Violative
    Establishments....................................226
    5.10.4.2 - Violative Establishments.........227
    5.10.4.3 - Individual Narrative
    Headings..227
5.10.5 - Exhibits.......................................240
    5.10.5.1 - Electronic information.............240
5.10.6 - Addendum To EIR..........................241

Part II

Exhibits, Appendix, and Samples............243

IOM Exhibits, Appendix, and Samples...........245
Exhibits...245
    Chapter 5 - Exhibits.........................245
Appendixes...246

Sample Schedules 246

IOM Chapter 5 Exhibits 249
Exhibit 5-1: Form FDA 482 Notice of
Inspection (2 Pgs) 250
Exhibit 5-2: FDA 482a Demand for Records 252
Exhibit 5-3: FDA 482b Request for
Information 253
Exhibit 5-4: Modified FDA 482 254
Exhibit 5-5: Form FDA 483 (2 Pgs) 255
Exhibit 5-6: Inserting Digital Photos into
Turbo EIR (Resize Photo) 256
Exhibit 5-7: Inserting Digital Photos into
Turbo EIR (Insert Photo) 256
Exhibit 5-8: Inserting Digital Photos into
Turbo EIR (Resize Using MS Office Picture
Manager) 256
Exhibit 5-9: Facts Create Assignment
Screen (1 Pg) 257
Exhibit 5-10: FDA 482(C) Notice of
Inspection Request for Records 258
Exhibit 5-11: Food Additives Nomographs 259
Exhibit 5-12: Summary of Registration and
Listing*** Human Pharmaceuticals (1 Pg) 261
Exhibit 5-13: Substantially Equivalent
Medical Devices (1 Pg) 262
Exhibit 5-14: Facts - Profile - Comstat
(6 Pgs) 266
Exhibit 5-15: Compliance Achievement
Report 266
Exhibit 5-16: Facts EI Record (3 Pgs) 267

Part III

Combined Glossary and Index 271

Combined Glossary 272

Index 287
About the author 337
About PharmaLogika 337
Other books available 339

Preface

About this Book

This book is first, and foremost, a unified reference source for the *Establishment Inspections* detailed in the United States Food and Drug Administration Investigations Operations Manual (Chapter 5). The procedures, guidance documents and some forms have also been collected. While the Investigations Operations Manual contains eight chapters and extensive addendums and samples, this text includes only the chapter and associated documents specific to *Establishment Inspections*.

The included *Overview and Orientation* (Chapter 2 of this book) is designed to provide a foundation for understanding the background of the FDA's regulations and FDA's relationship with the individual protectees.

This book was designed to be used both as a reference for experienced industry representatives and as a training resource for those new to the industry.

Included Documents and Features

- Investigations Operations Manual Chapter 5: Establishment Inspections

The IOM is the primary guidance document on FDA inspection policy and procedures for field investigators and inspectors.

- Exhibits, Appendix, and Sample Schedules

The exhibits support each chapter of the IOM with examples, tables, graphics, and forms. The exhibits available are listed in this book, but the published exhibits are available on the FDA's website.

Reference Tools

- Glossaries for each included part and procedure combined in one location

- Combined Index for all procedures and addendums to the Investigations Operations Manual

About the Reference Tools

Overview and Orientation

This overview provides the reader with a brief history of the FDA and explains not only what good manufacturing practice is, but why we have regulations and how they came to be.

The overview also lists all of the parts within Title 21 of the Code of Federal Regulations.

Combined Glossary

The Combined Glossary includes all of the definitions from each chapter listed alphabetically rather than by document.

When a word or term appears multiple times in the regulation and guidance documents, the word will appear multiple times in the Combined Glossary. Each duplicate entry is indented to highlight that it is a duplicate and the earliest reference to the entry is listed first. The source for each entry is bracketed (i.e., [Chapter 4]) for ease of reference. While the definitions are similar from one regulatory or guidance document to the next, they are not always identical.

Combined Index for all chapters and addendums

The index is composed of a list of both words and terms specific to investigations operations. It is a tool that assists in cross-referencing the various guidance (rather than having to rely on reading and comparing each chapter individually).

The index provides keywords and terminology as a tool to easily locate specific references across *all* chapters of and addendums to the Investigations Operations Manual rather than having to rely on memory or paging through the entire manual.

Overview and Orientation

The Food and Drug Administration (FDA)

The United States Food and Drug Administration (FDA) is responsible for protecting and promoting the nation's public health.

The programs for safety regulation vary widely by the type of product, its potential risks, and the regulatory powers granted to the agency. For example, the FDA regulates almost every facet of prescription drugs, including testing, manufacturing, labeling, advertising, marketing, efficacy and safety. FDA regulation of cosmetics, however, is focused primarily on labeling and safety. The FDA regulates most products with a set of published standards enforced by a combination of facility inspections, voluntary company reporting standards, and public and consumer watchdog activity.

The FDA frequently works in conjunction with other Federal agencies including the Department of Agriculture, Drug Enforcement Administration, Customs and Border Protection, and Consumer Product Safety Commission.

Historical Origins of Federal Food and Drug Regulation

Prior to the 20th century, there were few federal laws regulating the contents and sale of domestically produced food and pharmaceuticals before the 20th century (with one exception being the short-lived Vaccine Act of 1813). Some state laws provided varying degrees of protection against unethical sales practices, such as misrepresenting the ingredients of food products or therapeutic substances.

The history of the FDA can be traced to the latter part of the 19th century and the U.S. Department of Agriculture's Division of Chemistry (later Bureau of Chemistry). Under Harvey Washington Wiley, appointed chief chemist in 1883, the Division began conducting research into the adulteration and misbranding of food and drugs on the

American market. Although they had no regulatory powers, the Division published its findings from 1887 to 1902 in a ten-part series entitled Foods and Food Adulterants. Wiley used these findings, and alliances with diverse organizations (such as state regulators, the General Federation of Women's Clubs, and national associations of physicians and pharmacists) to lobby for a new federal law to set uniform standards for food and drugs to enter into interstate commerce.

Wiley's advocacy came at a time when the public had become alert to hazards in the marketplace by journalists and became part of a general trend for increased federal regulation in matters pertinent to public safety during the Progressive Era.[1] The 1902 Biologics Control Act was put in place after diphtheria antitoxin was collected from a horse named Jim who also had tetanus, resulting in several deaths.

The 1906 Food and Drug Act and creation of the FDA

In June 1906, President Theodore Roosevelt signed into law the Food and Drug Act, also known as the "Wiley Act" after its chief advocate.[2] The Act prohibited, under penalty of seizure of goods, the interstate transport of food which had been "adulterated," with that term referring to the addition of fillers of reduced "quality or strength," coloring to conceal "damage or inferiority," formulation with additives "injurious to health," or the use of "filthy, decomposed, or putrid" substances. The act applied similar penalties to the interstate marketing of "adulterated" drugs, in which the "standard of strength, quality, or purity" of the active ingredient was not either stated clearly on the label or listed in the United States Pharmacopoeia or the National Formulary. The act also banned "misbranding" of food and drugs.[3] The responsibility for examining food and drugs for such "adulteration" or "misbranding" was given to Wiley's USDA Bureau of Chemistry.[4] Strength, quality, identity, potency, and purity (SQuIPP) are currently the key product

---

[1] A History of the FDA at www.FDA.gov.
[2] A History of the FDA at www.FDA.gov.
[3] Text in quotation marks is the original text of the 1906 Food and Drugs Act and Amendments.
[4] A History of the FDA at www.FDA.gov.

---

safety standards, with only two measures added since 1906 Act.

Wiley used these new regulatory powers to pursue an aggressive campaign against the manufacturers of foods with chemical additives, but the Chemistry Bureau's authority was soon checked by judicial decisions, as well as by the creation of the Board of Food and Drug Inspection and the Referee Board of Consulting Scientific Experts as separate organizations within the USDA in 1907 and 1908 respectively. A 1911 Supreme Court decision ruled that the 1906 act did not apply to false claims of therapeutic efficacy,[5] in response to which a 1912 amendment added "false and fraudulent" claims of "curative or therapeutic effect" to the Act's definition of "misbranded." However, these powers continued to be narrowly defined by the courts, which set high standards for proof of fraudulent intent.[6] In 1927, the Bureau of Chemistry's regulatory powers were reorganized under a new USDA body, the Food, Drug, and Insecticide organization. This name was shortened to the Food and Drug Administration (FDA) three years later.[7]

The 1938 Food, Drug, and Cosmetic Act

By the 1930s, muckraking journalists, consumer protection organizations, and federal regulators began mounting a campaign for stronger regulatory authority by publicizing a list of injurious products which had been ruled permissible under the 1906 law, including radioactive beverages, cosmetics which caused blindness, and worthless "cures" for diabetes and tuberculosis. The resulting proposed law was unable to get through the Congress of the United States for five years, but was rapidly enacted into law following the public outcry over the 1937 Elixir Sulfanilamide tragedy, in which over 100 people died after using a drug formulated with a toxic, untested solvent. The only way that the FDA could even seize the product was due to a misbranding problem: an "Elixir" was defined as a medication dissolved in ethanol, not the diethylene glycol used in the Elixir Sulfanilamide.

---

[5] United States v. Johnson (31 S. Ct. 627 May 29, 1911, decided).

[6] A History of the FDA at www.FDA.gov.

[7] Milestones in U.S. Food and Drug Law History at www.FDA.gov.

President Franklin Delano Roosevelt signed the new Food, Drug, and Cosmetic Act (FD&C Act) into law on June 24, 1938. The new law significantly increased federal regulatory authority over drugs by mandating a pre-market review of the safety of all new drugs, as well as banning false therapeutic claims in drug labeling without requiring that the FDA prove fraudulent intent. The law also authorized factory inspections and expanded enforcement powers, set new regulatory standards for foods, and brought cosmetics and therapeutic devices under federal regulatory authority. This law, though extensively amended in subsequent years, remains the central foundation of FDA regulatory authority to the present day.[8]

Early FD&C Act amendments: 1938-1958

Soon after passage of the 1938 Act, the FDA began to designate certain drugs as safe for use only under the supervision of a medical professional, and the category of 'prescription-only' drugs was securely codified into law by the 1951 Durham-Humphrey Amendment.[9] While pre-market testing of drug efficacy was not authorized under the 1938 FD&C Act, subsequent amendments such as the Insulin Amendment and Penicillin Amendment did mandate potency testing for formulations of specific lifesaving pharmaceuticals.[10] The FDA began enforcing its new powers against drug manufacturers who could not substantiate the efficacy claims made for their drugs, and the United States Court of Appeals for the Ninth Circuit ruling in Alberty Food Products Co. v. United States (1950) found that drug manufacturers could not evade the "false therapeutic claims" provision of the 1938 act by simply omitting the intended use of a drug from the drug's label. These developments confirmed extensive powers for the FDA to enforce post-marketing recalls of ineffective drugs.[11] Much of the FDA's regulatory attentions in this era were directed towards abuse of amphetamines and barbiturates, but the agency also reviewed some 13,000 new drug applications between 1938 and 1962. While the science of toxicology was in its infancy at the start of this era, rapid advances in experimental assays for food additive and drug

---

[8] A History of the FDA at www.FDA.gov.
[9] A History of the FDA at www.FDA.gov.
[10] Milestones in U.S. Food and Drug Law History at www.FDA.gov
[11] A History of the FDA at www.FDA.gov.

---

safety testing were made during this period by FDA regulators and others.[12]

Good Manufacturing Practices vs. Current Good Manufacturing Practices

The terms "Good Manufacturing Practices (GMPs)" and "Current Good Manufacturing Practices (CGMPs or cGMPs[13])" are often used interchangeably both in industry and by FDA inspectors.

"Good Manufacturing Practices" generally refers to the legal mandates detailed in Title 21 of the Code of Federal Regulations (21CFR). "Current Good Manufacturing Practices" refers not only to the legal requirements, but to the guidance provided by the FDA and the standards practiced in industry that are not memorialized as law.

Organizational Structure

The FDA is an agency within the United States Department of Health and Human Services responsible for protecting and promoting the nation's public health. It is organized into the following major subdivisions, each focused on a major area of regulatory responsibility:

- The Office of the Commissioner (OC)
- The Center for Drug Evaluation and Research (CDER)
- The Center for Biologics Evaluation and Research (CBER)
- The Center for Food Safety and Applied Nutrition (CFSAN)
- The Center for Devices and Radiological Health (CDRH)
- The Center for Veterinary Medicine (CVM)
- The National Center for Toxicological Research (NCTR)
- The Office of Regulatory Affairs (ORA)
- The Office of Criminal Investigations (OCI)

---

[12] A History of the FDA at www.FDA.gov.
[13] The lower case "c" was coined in industry to differentiate between the law, emphasized with capital letters, and the current accepted industry practice not mandated by law.

How does the FDA communicate with Industry?

Code of Federal Regulations[14]

The Code of Federal Regulations (CFR) is the codification of the general and permanent rules and regulations (sometimes called administrative law) published in the Federal Register by the executive departments and agencies of the Federal Government of the United States. The CFR is published by the Office of the Federal Register, an agency of the National Archives and Records Administration (NARA).

The CFR is divided into 50 titles that represent broad areas subject to Federal regulation. Title 21 is the portion of the Code of Federal Regulations that governs food and drugs within the United States for the Food and Drug Administration (FDA), the Drug Enforcement Administration (DEA), and the Office of National Drug Control Policy (ONDCP).

It is divided into three chapters:

- Chapter I — Food and Drug Administration
- Chapter II — Drug Enforcement Administration
- Chapter III — Office of National Drug Control Policy

Guidance Documents

Guidance documents represent the Agency's current thinking on a particular subject. They do not create or confer any rights for or on any person and do not operate to bind FDA or the public. An alternative approach may be used if such approach satisfies the requirements of the applicable statute, regulations, or both. For information on a specific guidance document, please contact the originating office. Another method of obtaining guidance documents is through the Division of Drug Information.

---

[14] Available CFR Titles on GPO Access at http://www.access.gpo.gov/nara/cfr/cfr-table-search.html#page1

Federal Register

The Federal Register (since March 14, 1936), abbreviated FR, or sometimes Fed. Reg.) is the official journal of the federal government of the United States that contains most routine publications and public notices of government agencies. It is a daily (except holidays) publication.

The Federal Register is compiled by the Office of the Federal Register (within the National Archives and Records Administration) and is printed by the Government Printing Office.

There are no copyright restrictions on the Federal Register as it is a work of the U.S. government. It is in the public domain.[15]

Citations from the Federal Register are [volume] FR [page number] ([date]), e.g., 65 FR 741 (2000-10-01).

Direct Communication and Letters

The FDA interacts with consumers, health professionals, and industry representatives through letters, meetings (requested by either the FDA or the industry representatives), and telephone calls.

While not all questions can be answered over the phone, the FDA prefers telephone interactions over physical meetings (when a teleconference can reasonably replace a face-to-face meeting).

www.FDA.gov

The FDA maintains a website at www.fda.gov that is focused on three key audiences:

- consumers
- health professionals
- industry representatives

Through collaboration with users in testing site-wide designs, FDA.gov provides online access to its guidance documents, communication with industry, consumers, and

---

[15] The Federal Register at the GPO, online in both text and PDF, from 1994 on.

health professionals. Information is categorized by topic, with related subjects consolidated in sections on the site.

Additionally, FDA.gov provides a search engine for Title 21 of the CFR that makes finding keyword references throughout the title more accessible.

Conferences

The FDA routinely sends speakers to industry conferences where they are available to answer questions on their particular area of expertise.

False Statement to a Federal Agency

The U.S. Code of Federal Regulations (CFR) makes it a federal crime for anyone willfully and knowingly to make a false or fraudulent statement to a department or agency of the United States. The false statement must be related to a material matter, and the defendant must have acted willfully and with knowledge of the falsity. It is not necessary to show that the government agency was in fact deceived or misled. The issue of materiality is one of law for the courts. The maximum penalty is five years' imprisonment and a $10,000 fine.

A person may be guilty of a violation without proof that he or she had knowledge that the matter was within the jurisdiction of a federal agency. A businessperson may violate this law by making a false statement to another firm or person with knowledge that the information will be submitted to a government agency. Businesses must take care to avoid exaggerations in the context of any matter that may come within the jurisdiction of a federal agency.

CFR Title 21 - Food and Drugs: Parts 1 to 1499[16]

General

(1) General enforcement regulations

(2) General administrative rulings and decisions

---

[16] All of the 21CFR regulations can be searched online for no charge at http://www.accessdata.fda.gov/scripts/cdrh/cfdocs/cfcfr/cfrsearch.cfm

(3) Product jurisdiction

(5) Organization

(7) Enforcement policy

(10) Administrative practices and procedures

(11) Electronic records; electronic signatures

(12) Formal evidentiary public hearing

(13) Public hearing before a public board of inquiry

(14) Public hearing before a public advisory committee

(15) Public hearing before the commissioner

(16) Regulatory hearing before the food and drug administration

(17) Civil money penalties hearings

(19) Standards of conduct and conflicts of interest

(20) Public information

(21) Protection of privacy

(25) Environmental impact considerations

(26) Mutual recognition of pharmaceutical good manufacturing practice

(50) Protection of human subjects

(54) Financial disclosure by clinical investigators

(56) Institutional review boards

(58) Good laboratory practice for nonclinical laboratory studies

(60) Patent term restoration

(70) Color additives

(71) Color additive petitions

(73) Listing of color additives exempt from certification

(74) Listing of color additives subject to certification

(80) Color additive certification

(81) General specifications and general restrictions for provisional color additives for use in foods, drugs, and cosmetics

(82) Listing of certified provisionally listed colors and specifications

(83-98) [reserved]

(99) Dissemination of information on unapproved/new uses for marketed drugs, biologics, and devices

Foods

(100) General

(101) Food labeling

(102) Common or usual name for nonstandardized foods

(104) Nutritional quality guidelines for foods

(105) Foods for special dietary use

(106) Infant formula quality control procedures

(107) Infant formula

(108) Emergency permit control

(109) Unavoidable contaminants in food for human consumption and food-packaging material

(110) Current good manufacturing practice in manufacturing, packing, or holding human food

(111) Current good manufacturing practice in manufacturing, packaging, labeling, or holding operations for dietary supplements

(113) Thermally processed low-acid foods packaged in hermetically sealed containers

(114) Acidified foods

(115) Shell eggs

(119) Dietary supplements that present a significant or unreasonable risk

(120) Hazard analysis and critical control point (HACCP) systems

(123) Fish and fishery products

(129) Processing and bottling of bottled drinking water

(130) Food standards: general

(131) Milk and cream

(133) Cheeses and related cheese products

(135) Frozen desserts

(136) Bakery products

(137) Cereal flours and related products

(139) Macaroni and noodle products

(145) Canned fruits

(146) Canned fruit juices

(150) Fruit butters, jellies, preserves, and related products

(152) Fruit pies

(155) Canned vegetables

(156) Vegetable juices

(158) Frozen vegetables

(160) Eggs and egg products

(161) Fish and shellfish

(163) Cacao products

(164) Tree nut and peanut products

(165) Beverages

(166) Margarine

(168) Sweeteners and table syrups

(169) Food dressings and flavorings

(170) Food additives

(171) Food additive petitions

(172) Food additives permitted for direct addition to food for human consumption

(173) Secondary direct food additives permitted in food for human consumption

(174) Indirect food additives: general

(175) Indirect food additives: adhesives and components of coatings

(176) Indirect food additives: paper and paperboard components

(177) Indirect food additives: polymers

(178) Indirect food additives: adjuvants, production aids, and sanitizers

(179) Irradiation in the production, processing and handling of food

(180) Food additives permitted in food or in contact with food on an interim basis pending additional study

(181) Prior-sanctioned food ingredients

(182) Substances generally recognized as safe

(184) Direct food substances affirmed as generally recognized as safe

(186) Indirect food substances affirmed as generally recognized as safe

(189) Substances prohibited from use in human food

(190) Dietary supplements

(191-199) [reserved]

Drugs

(200) General

(201) Labeling

(202) Prescription drug advertising

(203) Prescription drug marketing

(205) Guidelines for state licensing of wholesale prescription drug distributors

(206) Imprinting of solid oral dosage form drug products for human use

(207) Registration of producers of drugs and listing of drugs in commercial distribution

(208) Medication guides for prescription drug products

(209) Requirement for authorized dispensers and pharmacies to distribute a side effects statement

(210) Current good manufacturing practice in manufacturing, processing, packing, or holding of drugs; general

(211) Current good manufacturing practice for finished pharmaceuticals

(216) Pharmacy compounding

(225) Current good manufacturing practice for medicated feeds

(226) Current good manufacturing practice for type A medicated articles

(250) Special requirements for specific human drugs

(290) Controlled drugs

(299) Drugs; official names and established names

New Drugs and Over-the-Counter Drug Products

(300) General

(310) New drugs

(312) Investigational new drug application

(314) Applications for FDA approval to market a new drug

(315) Diagnostic radiopharmaceuticals

(316) Orphan drugs

(320) Bioavailability and bioequivalence requirements

(328) Over-the-counter drug products intended for oral ingestion that contain alcohol

(330) Over-the-counter (OTC) human drugs which are generally recognized as safe and effective and not misbranded

(331) Antacid products for over-the-counter (OTC) human use

(332) Antiflatulent products for over-the-counter human use

(333) Topical antimicrobial drug products for over-the-counter recognized as safe and effective and not misbranded

(335) Antidiarrheal drug products for over-the-counter human use

(336) Antiemetic drug products for over-the-counter human use

(338) Nighttime sleep-aid drug products for over-the-counter human use

(340) Stimulant drug products for over-the-counter human use

(341) Cold, cough, allergy, bronchodilator, and antiasthmatic drug products for over-the-counter human use

(343) Internal analgesic, antipyretic, and antirheumatic drug products for over-the-counter human use

(344) Topical OTIC drug products for over-the-counter human use

(346) Anorectal drug products for over-the-counter human use

(347) Skin protectant drug products for over-the-counter human use

(348) External analgesic drug products for over-the-counter human use

(349) Ophthalmic drug products for over-the-counter human use

(350) Antiperspirant drug products for over-the-counter human use

(352) Sunscreen drug products for over-the-counter human use [stayed indefinitely]

(355) Anticaries drug products for over-the-counter human use

(357) Miscellaneous internal drug products for over-the-counter human use

(358) Miscellaneous external drug products for over-the-counter human use

(361) Prescription drugs for human use generally recognized as safe and effective and not misbranded: drugs used in research

(369) Interpretative statements re warnings on drugs and devices for over-the-counter sale

(370-499) [reserved]

Veterinary Products

(500) General

(501) Animal food labeling

---

(502) Common or usual names for nonstandardized animal foods

(509) Unavoidable contaminants in animal food and food-packaging material

(510) New animal drugs

(511) New animal drugs for investigational use

(514) New animal drug applications

(515) Medicated feed mill license

(516) New animal drugs for minor use and minor species

(520) Oral dosage form new animal drugs

(522) Implantation or injectable dosage form new animal drugs

(524) Ophthalmic and topical dosage form new animal drugs

(526) Intramammary dosage forms

(529) Certain other dosage form new animal drugs

(530) Extralabel drug use in animals

(556) Tolerances for residues of new animal drugs in food

(558) New animal drugs for use in animal feeds

(564) [reserved]

(570) Food additives

(571) Food additive petitions

(573) Food additives permitted in feed and drinking water of animals

(579) Irradiation in the production, processing, and handling of animal feed and pet food

(582) Substances generally recognized as safe

(584) Food substances affirmed as generally recognized as safe in feed and drinking water of animals

(589) Substances prohibited from use in animal food or feed

(590-599) [reserved]

Biologics

(600) Biological products: general

(601) Licensing

(606) Current good manufacturing practice for blood and blood components

(607) Establishment registration and product listing for manufacturers of human blood and blood products

(610) General biological products standards

(630) General requirements for blood, blood components, and blood components, and blood derivatives

(640) Additional standards for human blood and blood products

(660) Additional standards for diagnostic substances for laboratory tests

(680) Additional standards for miscellaneous products

Cosmetics

(700) General

(701) Cosmetic labeling

(710) Voluntary registration of cosmetic product establishments

(720) Voluntary filing of cosmetic product ingredient composition statements

(740) Cosmetic product warning statements

(741-799) [reserved]

Medical Devices

(800) General

(801) Labeling

(803) Medical device reporting

(806) Medical devices; reports of corrections and removals

(807) Establishment registration and device listing for manufacturers and initial importers of devices

(808) Exemptions from federal preemption of state and local medical device requirements

(809) In vitro diagnostic products for human use

(810) Medical device recall authority

(812) Investigational device exemptions

(813) [reserved]

(814) Premarket approval of medical devices

(820) Quality system regulation

(821) Medical device tracking requirements

(822) Postmarket surveillance

(860) Medical device classification procedures

(861) Procedures for performance standards development

(862) Clinical chemistry and clinical toxicology devices

(864) Hematology and pathology devices

(866) Immunology and microbiology devices

(868) Anesthesiology devices

(870) Cardiovascular devices

(872) Dental devices

(874) Ear, nose, and throat devices

(876) Gastroenterology-urology devices

(878) General and plastic surgery devices

(880) General hospital and personal use devices

(882) Neurological devices

(884) Obstetrical and gynecological devices

(886) Ophthalmic devices

(888) Orthopedic devices

(890) Physical medicine devices

(892) Radiology devices

(895) Banned devices

(898) Performance standard for electrode lead wires and patient cables

Mammography

(900) Mammography

Radiological Health

(1000) General

(1002) Records and reports

(1003) Notification of defects or failure to comply

(1004) Repurchase, repairs, or replacement of electronic products

(1005) Importation of electronic products

(1010) Performance standards for electronic products: general

(1020) Performance standards for ionizing radiation emitting products

(1030) Performance standards for microwave and radio frequency emitting products

(1040) Performance standards for light-emitting products

(1050) Performance standards for sonic, infrasonic, and ultrasonic radiation-emitting products

Regulations under Certain Other Acts

(1210) Regulations under the federal import milk act

(1230) Regulations under the federal caustic poison act

(1240) Control of communicable diseases

(1250) Interstate conveyance sanitation

(1251-1269) [reserved]

(1270) Human tissue intended for transplantation

(1271) Human cells, tissues, and cellular and tissue-based products

(1272-1299) [reserved]

Controlled Substances

(1300) Definitions

(1301) Registration of manufacturers, distributors, and dispensers of controlled substances

(1302) Labeling and packaging requirements for controlled substances

(1303) Quotas

(1304) Records and reports of registrants

(1305) Orders for schedule I and II controlled substances

(1306) Prescriptions

(1307) Miscellaneous

(1308) Schedules of controlled substances

(1309) Registration of manufacturers, distributors, importers and exporters of List I chemicals

(1310) Records and reports of listed chemicals and certain machines

(1311) Digital certificates

(1312) Importation and exportation of controlled substances

(1313) Importation and exportation of list I and list II chemicals

(1314) Retail sale of scheduled listed chemical products

(1315) Importation and production quotas for ephedrine, pseudoephedrine, and phenylpropanolamine

(1316) Administrative functions, practices, and procedures

Office of National Drug Control Policy

(1400) [reserved]

(1401) Public availability of information

(1402) Mandatory declassification review

(1403) Uniform administrative requirements for grants and cooperative agreements to state and local governments

(1404) Governmentwide debarment and suspension (nonprocurement)

(1405) Governmentwide requirements for drug-free workplace (financial assistance)

(1406-1499) [reserved]

Part I

Investigations Operations Manual

Foreword

Last Update / Version: June 18, 2009

The *Investigations Operations Manual* (IOM) is the primary source regarding Agency policy and procedures for field investigators and inspectors. This extends to all individuals who perform field investigational activities in support of the Agency's public mission. Accordingly, it directs the conduct of all fundamental field investigational activities. Adherence to this manual is paramount to assure quality, consistency, and efficiency in field operations. The specific information in this manual is supplemented, not superseded, by other manuals and field guidance documents. Recognizing that this manual may not cover all situations or variables arising from field operations, any significant departures from IOM established procedures should have the concurrence of district management.

For 2009, the IOM contains some important changes which clarify or present new procedures.For instance, the LACF/AF Sample Schedule was revised and sample identification procedures were clarified in Chapter 4; additional FDA 483 guidance was added; OEI maintenance instructions reference FMD-130; and additional information related to food, drug, and veterinary products was incorporated in Chapter 5; and current import terminology was updated in Chapter6. Appendix C - Blood Values was completely revised.

As with each new edition of the IOM, please take time to review sections of the IOM for changes which may apply to your work.

Since December 1996, the IOM has been posted on ORA's Internet Website, Investigations Operations Manual. The entire IOM is available there, with all graphics included. Future updates to the IOM will be performed periodically during the year to this on-line version. The hard copy will be published annually. Remember, whether reviewing the "hard copy" or the "on-line' version of the IOM, the most recent version is the document of record.

We are committed to the continual improvement of the quality and usefulness of the IOM. Suggestions for the 2009 edition of the IOM or recommended changes, deletions, additions to the IOM may be sent to the Division of Field Investigations (HFC-130), 5600 Fishers Lane, Rockville, MD 20857 or via e-mail to the Director, DFI. You can also e-mail IOM@FDA.HHS.GOV. If you are recommending a change or revision, please use the IOM Change Request Form available from the web site.

Thank you for your continued hard work and dedication in protecting and promoting the health and well-being of the American people.

Michael A. Chappell, Acting Associate Commissioner for Regulatory Affairs

FDA/Office of Regulatory Affairs

Note: *This manual is reference material for investigators and other FDA personnel. The document does not bind FDA and does not confer any rights, privileges, benefits or immunities for or on any person(s).*

Vision / Mission / Values

Vision

All food is safe; all medical products are safe and effective; and the public health is protected and advanced.

Mission

Protecting consumers and enhances public health by maximizing compliance of FDA-regulated products and minimizing risk associated with those products.

Values

Values that I believe exist in ORA today:

1.   Our mission defines our work and our purpose as employees.

2.   We are dedicated to having a meaningful and positive impact on public health.

3. We take pride in our ongoing commitment to excellence in our work products and our contribution to the mission of FDA.

Values that I believe we need for the future:

1. We work as a team and build partnerships with our stakeholders in a global regulatory environment.

2. We invest in our people and processes to target and improve our impact on public health.

3. We continuously improve both what we do and how we do it.

Michael A. Chappell

Acting Associate Commissioner for Regulatory Affairs

December 9, 2008

Investigations Operations Manual Chapter 5: Establishment Inspections

CONTENT		VERSION DATE
Subchapter 5.1	Inspection Information	June 11, 2009
Subchapter 5.2	Inspection Procedures	June 11, 2009
Subchapter 5.3	Evidence Development	June 11, 2009
Subchapter 5.4	Food	June 11, 2009
Subchapter 5.5	Drugs	June 11, 2009
Subchapter 5.6	Devices	August 27, 2009
Subchapter 5.7	Biologics	June 11, 2009
Subchapter 5.8	Pesticides	June 11, 2009
Subchapter 5.9	Veterinary Medicine	June 11, 2009
Subchapter 5.10	Reporting	June 11, 2009

5.1 - Inspection Information

5.1.1 - Authority to Enter and Inspect

See IOM 2.2 for discussion of statutory authority.

It is your obligation to fulfill the following requirements because failure to do so may prevent use of evidence and information obtained during the inspection.

There may be occasions where you may be accompanied on your inspection or investigation by other officials. These officials may be state or local officials who have their own inspectional authority or other officials who do not have authority to enter the firm. You should obtain permission from the firm's most responsible person if officials without inspection authority wish to accompany you during your inspection/investigation. You should document in your EIR when other non-FDA officials accompany you during your inspection, and whether they entered under their own authority or the responsible individual at the firm gave permission (identify, by name and title, the responsible individual giving permission). See IOM 5.2.2 and 5.10.4.3.2.

5.1.1.1 - FDA Investigator's Responsibility

Your authority to enter and inspect establishments is predicated upon specific obligations to the firm as described below. It is your responsibility to conduct all inspections at reasonable times and within reasonable limits and in a reasonable manner. Proceed with diplomacy, tact and persuasiveness.

5.1.1.2 - Credentials

Display your credentials to the top management official be it the owner, operator, or agent in charge. See IOM 5.2.2.

NOTE: *Although management may examine your credentials and record the number and your name, do not permit your credentials to be photocopied. Federal Law (Title 18, U.S.C. 701) prohibits photographing, counterfeiting, or misuse of official credentials.*

5.1.1.3 - Written Notice

After showing the firm's representative your credentials, issue the original, properly executed, and signed FDA 482, Notice of Inspection, to the top management official. Keep the carbon copy for submission with your report.

5.1.1.4 - Written Observations

Upon completing the inspection and before leaving the premises, provide the highest management official available your inspectional findings on an FDA 483 - Inspectional Observations. See Section 704(b) of the FD&C Act [21 U.S.C. 374 (b)] and IOM 5.2.3 and 5.2.7.

5.1.1.5 - Receipts

Furnish the top management official the original of the FDA-484 - Receipt for Samples describing any samples obtained during the inspection. See IOM 5.2.4.

5.1.1.6 - Written Demand for Records

In low-acid canned food and acidified food EI's, an FDA 482a - Demand for Records (exhibit 5-2) is required under 21 CFR 108.35(h) and 21 CFR 108.25(g) to obtain records required by 21 CFR 113 and 114.

5.1.1.7 - Written Requests for Information

There are several methods of requesting records. These may include a request for information under LACF or AF inspections, 703 written requests, and requests for records under the BT Act (IOM 5.4.1.3).

5.1.1.7.1 - LACF / AF Food Inspections

In low-acid canned foods and acidified foods EIs, an FDA 482b, Request for Information (exhibit 5-3), is required under 21 CFR 108.35(c)(3)(ii) and 21 CFR 108.25(c)(3)(ii) to obtain information concerning processes and procedures required under 21 CFR 113 and 114.

5.1.1.7.2 - Requests for Records Under Section 703 of the FD&C Act

Per CPG Sec. 160.300, Requests for Records under Section 703 [21 U.S.C. 373], evidence obtained in response to a specific written request under Section 703 cannot be used in a criminal prosecution of the person from whom obtained. With Supervisory approval, in certain circumstances, you may decide to issue a 703 written request when the importance of the evidence is crucial to protecting the public health.

Procedure: All 703 written requests must comply with IOM 4.4.7.2.2. Consider obtaining the evidence from other sources before using the 703 written request. In the case of foods and feeds, if there is a risk or threat of serious adverse health consequences, the district should invoke the BT Act records access authority. All BT Act records requests must comply with IOM 5.4.1.3.

5.1.1.8 - Business Premises

Authority to inspect firms operating at a business location is described in IOM 5.1.1 and requires issuing management an FDA 482, Notice of Inspection, and presenting your credentials. A warrant for inspection is not necessary unless a refusal or partial refusal is encountered or anticipated.

5.1.1.9 - Premises Used for Living Quarters

All inspections where the premises are also used for living quarters must be conducted with a warrant for inspection unless:

Owner Agreeable - The owner or operator is fully agreeable and offers no resistance or objection whatsoever or;

Physically Separated - The actual business operations to be inspected are physically separated from the living quarters by doors or other building construction. These would provide a distinct division of the premises into two physical areas, one for living quarters and the other for business operations, and you do not enter the living area.

In both the latter cases, proceed as any other inspection with the appropriate presentation of credentials and issuance of a Notice of Inspection.

5.1.1.10 - Facilities where Electronic Products are Used or Held

Section 537(a) of the FD&C Act provides the FDA with the authority to inspect the facilities of manufacturers in certain circumstances. The electronic product radiation control provisions were originally enacted as the Radiation Control for Health and Safety Act of 1968 (P.L. 90-602)

It is lawful for FDA personnel to enter the facilities of an electronic product distributor, dealer, assembler or user for the purpose of testing an electronic product for radiation safety when the entry is voluntarily permitted. Congress has not specifically prohibited FDA from conducting such voluntary examinations and such examinations would clearly agree with the congressional declaration of purpose expressed in section 532(a) of the RCH&S Act.

Under the Medical Device Authority, electronic products utilized in human and/or veterinary medicine, e.g., x-ray, laser, ultra-sound, diathermy, etc. can be considered prescription devices. In these cases the authority of Section 704 of the FD&C Act [21 U.S.C. 374] can be used to obtain entry to inspect the user facility. If the Medical Device Authority is utilized, credentials must be displayed and a FDA 482, Notice of Inspection, must be issued.

5.1.1.11 - Multiple Occupancy Inspections

You are required per FD&C Act 704(a)(1) [21 U.S.C. 374(a)(1)] to issue a Notice of Inspection, FDA 482, to each firm inspected. When firms have operations located in different sites or buildings, you should use judgment to determine when multiple FDA 482 forms need to be issued. For sites located a distance apart, it is preferable to issue a FDA 482 to the most responsible person at each site. One rule of thumb which can be used is if the sites or buildings are within walking distance, your original Notice of Inspection can be considered sufficient to cover both. During your initial interview with management, after you issue the FDA 482, make sure you clearly indicate the facility and sites you intend to inspect. The Act requires the issuance of a Notice of Inspection, but does not prohibit issuing multiple notices if management so requests. As with all of our work, good judgment, and knowledge of the OEI and the FD&C Act are necessary in deciding what legally must be done.

5.1.1.12 - Authority for Examinations and Investigations

Section 702(a) of the FD&C Act [21 U.S.C. 372 (a)] authorizes examinations and investigations for the purpose of enforcing the Act.

5.1.1.13 - Authority to Implement Section 702(e)(5) of the FD&C Act

Section 702(e) of the FD&C Act [21 U.S.C. 372 (e)] contains certain authorities relating to counterfeit drugs including the authority to seize ("confiscate") counterfeit drugs and containers, counterfeiting equipment, and all other items used or designed for use in making counterfeit drugs prior to the initiation of libel proceedings. This authority has been delegated, with certain restrictions, to holders of official credentials consistent with their authority to conduct enforcement activities. Additional authority in 702(e) to make arrests, to execute and serve arrest warrants, to carry firearms, or to execute seizure by process under Section 304 of the FD&C Act [21 U.S.C. 334] have not been delegated.

The agency does intend to utilize the authority contained in Section 702(e) to execute and serve search warrants, but such use does not require delegation from the ACRA.

Section 702(e)(5) contains authority for such delegated persons to confiscate all items which are, or which the investigator has reasonable grounds to believe are, subject to seizure under Section 304(a)(2). Items subject to seizure, and thus to confiscation under Section 702(e)(5), includes most things associated with counterfeit drugs. Confiscation authority does not, however, extend to vehicles, records, or items (i.e., the profits) obtained as a result of counterfeiting.

5.1.1.13.1 - Scope

Under this delegation, with supervisory concurrence and prior to the initiation of libel proceedings, investigators and inspectors are authorized to confiscate:

1. Any counterfeit drug

2. Any container used to hold a counterfeit drug,

3. Any raw material used in making a counterfeit drug

4. Any labeling used for counterfeit drug

5. Any equipment used to make a counterfeit drug including punches, dies, plates, stones, tableting machines, etc.

6. Any other thing which you have reasonable grounds to believe is designed or used in making a counterfeit drug.

NOTE: *You and your supervisor must be constantly aware of the potential dangers involved in confiscating property from individuals. Special care should be taken to ensure your safety. Arranging for teams of investigators to conduct the investigation, or arranging for assistance by local police, or other agencies with police powers, should be considered in planning the confiscation of counterfeit materials.*

5.1.1.13.2 - Inspectional Guidance

Guidance provided for implementing the authority to confiscate drug counterfeits is as follows:

1. The authority is not to be utilized unless there has been an agency determination the drug to be confiscated is a counterfeit and it is a drug which "without authorization, bears a trademark, *** or any likeness" of a legitimate product. The determination usually is based upon evidence supplied by the firm whose product is being counterfeited. A written agency determination will issue to the District Director from the Office of Enforcement, in conjunction with the Office of Regional Operations, and the Center for Drug Evaluation and Research.

2. When engaged in counterfeit investigations, you should proceed as follows upon encountering items to be confiscated.

   a. Evaluate safety needs and check the location to ensure it is safe to proceed. Do not attempt to remove an item by force. If it appears there will be resistance, contact the local police, or other agencies with police powers for backup, if not already done in advance.

   b. Inventory the items to be confiscated.

   c. Prepare a written receipt and offer it to the person in charge.

   d. Remove the items, if possible, from the premises (if they cannot be removed, secure them under seal).

   e. Place all items removed under lock at a secure location. In most cases, confiscated items will be stored at the district or resident post office until they are seized.

5.1.1.13.3 - Follow up Guidance

After items are confiscated, certain actions must be taken to bring confiscated items under the control of the court. Proceed as follows:

1. After an item is confiscated, immediately notify your supervisor.

2. Supervisors must then notify the appropriate compliance units of the items confiscated.

3. Compliance units should initiate seizure proceedings against any items confiscated.

4. ORO/DFI should be advised of any action utilizing this authority.

5.1.1.13.4 - Search Warrants

Section 702(e)(2) contains authority to execute and serve search warrants. Proceed as instructed by your district after a search warrant has been obtained.

5.1.1.14 - Products Imported Under the Provisions of Section 801(d)(3) of the FD&C Act

The FDA Export Reform and Enhancement Act of 1996 (PL 104-134 and 104-180) amended the FD&C Act by adding Section 801(d)(3) ("Import for Export") which permits the importation of unapproved drug and medical device components, food additives, color additives, and dietary supplements intended for further incorporation or processing into products destined for export from the United States. Section 801(d)(3) was subsequently amended by Section 322 of the Public Health Security and Bioterrorism Preparedness and Response Act of 2002 (Bioterrorism Act), Public Law 107-188, which specified certain requirements an importer has to satisfy in order to import a product under this Section. See IOM 6.2.3.4.

5.1.1.14.1 - Requirements for BT Act

These requirements include:

1. A statement confirming the intent to further process such article or incorporate such article into a product to be exported,

2. The identification of all entities in the chain of possession of the imported article,

3. A certificate of analysis "as necessary to identify the article" (unless the article is a device), and

4. Executing a bond providing for liquidated damages in the event of default, in accordance with U.S. Customs. This bond remains in effect until the final product is exported and destroyed.

In addition, the initial owner or consignee must keep records showing the use of the imported articles, and must be able to provide upon request a report showing the disposition or export of the imported articles. An article imported under this section, and not incorporated or further processed, must be destroyed or exported by the owner or consignee. Failure to keep records or to make them available to FDA, making false statements in such records, failure to export or destroy imported articles not further incorporated into finished products, and introduction of the imported article or final product into domestic commerce are Prohibited Acts under Section 301(w).

Filers making entry under the Import for Export provisions must either identify entry submissions with the OASIS Affirmation of Compliance "IFE" (Import for Export), or supply FDA with written documentation stating the product is entered under the Import for Export provisions. A Certificate of Analysis (as necessary) and identification of all involved entities must be submitted in writing to the import district. The import district will forward all written documentation to the home district of the initial owner or consignee for incorporation into the appropriate Establishment File.

5.1.1.14.2 - Inspectional Preparation

Before conducting an Establishment Inspection, contact your district's designated individual with access to OASIS/EEPS Reports to obtain a printout of any import entries made by the establishment under the Import for Export provisions through OASIS. In addition, check the district factory file for copies of any Import for Export documents forwarded from the district where entry was filed. During the inspection examine the firm's records to determine the disposition of any items identified at time of entry as intended for incorporation into products for export. Document any instances in which such products were introduced into domestic commerce or cannot be accounted for (see IOM 6.2.3.4.3).

5.1.2 - Inspectional Approach

An establishment inspection is a careful, critical, official examination of a facility to determine its compliance with laws administered by FDA. Inspections may be used to obtain evidence to support legal action when violations are found, or they may be directed to obtaining specific information on new technologies, good commercial practices, or data for establishing food standards or other regulations. In order to facilitate on-the-job training, multiple points of view, and perspectives of firms being inspected whenever practical, those with assignment authority, should consider assigning different Investigator/s or different Lead Investigators at different times. This is recommended particularly when there have been multiple sequential NAI inspections or when the firm's management has been uncooperative.

The kind and type of inspection you conduct will normally be defined by the program, assignment, or your supervisor; according to the following definitions:

Comprehensive Inspection - directs coverage to everything in the firm subject to FDA jurisdiction to determine the firms compliance status; or

Directed Inspection - directs coverage to specific areas to the depth described in the program, assignment, or as instructed by your supervisor.

See IOM Subchapter 1.5 for information on safety, use of protective gear, dealing with potential hazards and other safety issues.

5.1.2.1 - Depth of Inspection

The degree and depth of attention given various operations in a firm depends upon information desired, or upon the violations suspected or likely to be encountered. In determining the amount of attention to be given in specific cases, consider:

1. The current Compliance Program,

2. Nature of the assignment,

3. General knowledge of the industry and its problems,

4. Firm history, and

---

5. Conditions found as the inspection progresses.

5.1.2.2 - Inspection Walk Through

A preliminary tour of the premises should be conducted early in an establishment inspection to become familiar with the operation and to plan the inspection strategy. A walk through visual inspection of the manufacturing site is helpful in establishing the depth of the inspection, learning about products and processes, identifying sources of manufacturing records and identifying potential areas of concern. The size of the facility, the number of employees, employee practices, environmental conditions inside and outside the plant, raw materials, manual and automated processes, sources of contamination, manufacturing flow, method of data collection including computer terminals, are some of the areas to be taken into consideration in establishing the depth of the inspection. A visual inspection of a manufacturing site should also be used to check obvious potential problem areas such as: general housekeeping, state of operation for processes and processing equipment, and people dependent operations. Visual inspections of areas used for failure investigation, product sampling and testing, product reworks, return goods, and product quarantine areas should be inspected for obvious potential product problems.

Depending on the product is being inspected, some of the general inspectional equipment an investigator should have available, may include, eye and ear protection, boots and protective clothing. Some specialized equipment may include radiation or EO monitoring devices, magnifiers, and timing devices as needed. For some domestic and foreign plant sites, investigators may be required to be inoculated prior to the inspection for protection from potential environmental concerns such as hepatitis, yellow fever, malaria and live biological products which may be encountered in vaccine products. See subchapter IOM 1.5.

5.1.2.3 - Signing Non-FDA Documents

Occasionally a firm will request you sign various documents including:

1. A waiver which will exempt the firm from any responsibility or liability should an accident occur and you are injured on the firm's premises,

2. Form letters concerning access to confidential information the firm does not want released,

3. Information/data you request during the inspection be put into writing, etc.

If you receive such a request, inform the firm you are not authorized to sign such documents, letters, requests, waivers, etc., but will report the firm's request in your EIR. The use of common sense is expected with this procedure. All FDA employees are authorized to sign-in and sign-out at a firm and to comply with security measures employed by the firm, including documenting the removal/replacement of seals to inspect vehicles and containers. See IOM 4.3.4.3 and 4.5.4.6. Obviously, the key issue is you are not authorized to waive, without supervisory approval, any of FDA's rights to inspect, sample, photograph, copy, etc. or to sign any interstate shipping record document which could infer the firm could not be prosecuted under the Act.

5.1.2.4 - Technical Assistance

If you determine specialized technical assistance is necessary in conducting inspections of new technologies, products or manufacturing procedures, it may be available through Regional or National experts, other ORA components or Center scientists and engineers. If specialized skills are necessary and are not available locally or through your Region, contact the Division of Field Investigations (DFI), (HFC-130) at 301-827-5653. See FMD 142 and IOM 1.9.2.2.1 for additional information.

5.1.2.5 - Team Inspections

The use of teams to conduct inspections may be beneficial. Very often individuals well versed in an analytical or inspectional technique or technology can provide assistance and advice.

When inspection teams are involved in an inspection, one investigator will be designated as the team leader by the inspecting district or by the Division of Field Investigations (DFI/HFC-130) if a headquarters directed special inspection is involved. The team leader is in charge of the inspection and bears the overall responsibility for the inspection and the EIR. A team may consist of multiple investigators, laboratory personnel and other FDA employees, and your supervisor/coach, who may participate as part of the ORA Quality Assurance program.

5.1.2.5.1 - Team Member Responsibilities

Each team member is responsible for preparing those portions of the report pertaining to his/her activities. Team members shall identify their portion of the report so they can later identify that portion as the part he/she performed and reported. Since reports should be written in the first person, one system might be to head each portion with a statement "The following operation(s) was/were observed and reported by Investigator _____", who can then report in the first person.

All team members must sign the original EIR. Ideally, all team members should sign the FDA 483, if one is issued. However, issuance of the FDA 483 should not be delayed, in the absence of a team member's signature. See IOM 5.2.3 for instructions for signing a multi-page FDA 483.

5.1.2.5.2 - Team Leader Responsibilities

The Team Leader shall be responsible for:

1. Issuing unused notebooks for taking regulatory notes during the EI or investigation to headquarters personnel on the team. He/she is also responsible for instructions on their use, if necessary, and when the report is finished, for obtaining the headquarters individual's signature on the original EIR and completed and properly identified regulatory notes and submitting them to the supervisor for filing. See IOM 2.1.3.

2. Directing the overall inspection to accomplish the objectives of the assignment including;

    a. Planning the inspection,

    b. Scheduling and coordinating team members' pre-inspection preparations,

    c. Determining, to the extent possible, the firm will be open and operating,

    d. Planning for needs of visiting scientists if applicable. When the team leader is not familiar with all the processes or technology involved in the inspection, provide for primary coverage of selected areas by other team members,

e. Determining an orderly, efficient, and effective approach and sequence to be used and discussing the inspection plan with the team,

f. Modifying the inspection plan as necessary during the EI, to permit following leads, documenting evidence, etc.,

g. Setting team policy on how communications with the firm are to be handled,

h. Discussing personal conduct in dealing with headquarters personnel as necessary,

i. Assuring an early understanding by team members of their roles in note taking and reporting,

j. Assuring communications are open among team members, especially if the team is allowed to separate and work independently,

k. Reviewing inspection progress at least daily, discussing remaining objectives with the team members, and setting objectives for the following day,

l. Continually assessing the progress of the inspection to evaluate how the inspectional approach is working and to keep the district supervisor advised of the inspection's progress,

m. Providing guidance and direction to team members as necessary,

n. Advising each team member of reporting responsibilities and dates when drafts are to be provided,

o. Following up promptly on any delays or failures to report as required, and

p. Assisting the supervisor with further follow up, as indicated.

3. Making sure any person who joins the team after the inspection has started presents credentials and issues a Notice of Inspection, FDA 482, to the firm prior to actually taking part in the EI;

4. Completing and/or correcting the computer generated coversheet;

5. Preparing the Summary of Findings;

6. Completing all headings of an administrative nature in the narrative report;

7. Compiling and submitting the complete final report; and

8. Resolving any disputes or differences of opinion among the team members, including items, which may be listed on the FDA 483.

5.1.3 - Inspection of Foreign Firms

Inspectional requirements apply to all inspections, including foreign inspections. However, there are some exceptions. For instance the FDA 482 is not required, unless the firm is a US Military facility. Be guided by relevant Compliance Programs and the Guide to International Inspections and Travel Manual for other differences. See IOM 1.2.1.2.

5.1.4 - Inspectional Precautions

Our concern over microbiological contamination emphasizes the need for you to be alert to criticism or allegations that you may have contributed to or caused contamination at a firm. This is especially important in drug firms and high-risk food firms, among others. You must adhere to good sanitation practices to refute any such criticisms. You could also unknowingly introduce or spread disease during inspections of or visits to animal production or sale facilities, conducting environmental investigations at poultry layer facilities, conducting dairy farm inspections or audits of state activities, investigating tissue residue reports or working in the veterinary bioresearch area. See IOM 5.2.10 for information outlining precautions for you to follow.

Exercise caution in all activities in the firm. Follow the firm's sanitation program for employees and wash and sanitize hands, shoes, vehicles and equipment as indicated. Restrict unnecessary movement between various areas in plants and when possible, complete your activities in one area before moving to the next.

When inspecting areas where sterility is maintained or sterile rooms are located (especially in pharmaceutical or device firms), follow the sterile program required of the firm's employees. In general it is unnecessary to enter sterile rooms except in the most extraordinary circumstances. These areas are usually constructed to provide visual monitoring. Take no unsterile items with you (notebook, pencils, etc.). In this type of situation you can enter your observations in your regulatory notes immediately after leaving the sterile area.

Always use aseptic techniques, including hand sanitizing, when collecting in-line and raw material samples, as well as finished product samples for microbiological examination. See IOM 4.3.6.

Do not use or consume a firm's products at any of a firm's facilities. This could be interpreted as accepting a product as being satisfactory and could possibly embarrass you and the Agency, both during the inspection and in the future. In general, consuming food products in a manufacturing area is considered an objectionable practice.

When conducting inspections of firm's using chemicals, pesticides, etc., ask to review the Material Safety Data Sheets (MSDS) for the products involved to determine what, if any, safety precautions you must take. This could include the use of respirators or other safety equipment.

5.1.4.1 - Clothing

Wear clean coveralls or other protective clothing for each inspection and if circumstances dictate, use a clean pair when returning from lunch, or upon entering certain machinery or critical areas.

Remove and secure all jewelry, pens, pencils, notebook, etc., so they cannot fall into the product or machinery. Do not depend on clips on pens, etc., to hold these items in your outer pockets.

Clean protective clothing should be either individually wrapped or placed in clean plastic bags and taped to protect from contamination. If the package has been sterilized, protect the package from possible contamination or puncture. The package should not be opened until you are ready to use the clothing. After use, clothing should be turned inside out as it is removed, and immediately placed in clean paper or plastic bags to prevent spread of contamination until washed and/or sterilized.

Use disposable hair and head coverings throughout the inspection and disposable hand and foot coverings in areas where floor tracking or cross contamination may be a factor. Use hard hats and other protective devices where the situation dictates.

If reusable protective boots are used, wash and sanitize before each use. Always use sterile disposable boot covers when entering machinery such as dryers or where unavoidable contact with product is a factor.

When discarding contaminated disposable head and boot coverings, it is suggested they be placed with used clothing for proper disposal after leaving the plant area.

See IOM 5.2.10.1 for protective clothing and equipment necessary when visiting livestock or poultry producing areas.

5.1.4.2 - PHS Recommendations - Basic Sanitary Practices

FDA personnel are not required by law to have health certificates, take physical exams or submit to requirements, which ensures their compliance with sanitary procedures in the performance of their official duties. However, it is critical you adhere to basic sanitation practices. See IOM 1.5.1.5.

The Food Code is available electronically form the FDA CFSAN web page at http://www.cfsan.fda.gov under Federal/State Programs-Retail Food Safety References. Printed copies may be ordered from the National Technical Information Service, NTIS, www.ntis.gov.

5.1.4.3 - Representatives Invited by the Firm to View the Inspection

While conducting an inspection, you may find the firm's management has invited individuals who are not directly employed by the firm to view the inspectional process (e.g., representatives from the press, trade associations, consumer groups, congressional staff, other company officials).

Regardless of whom the firm invites to observe the progress of an inspection, the presence of outside representatives should not disrupt the inspectional process. You should continue to conduct the inspection in a reasonable fashion. The presence of these individuals should have no impact on the manner in which the inspection progresses except you should take precautions to preserve the confidentially of any information you may have obtained as a result of the Agency's statutory authority. This is especially true when the inspection is recorded via videotaping, other photography, and/or audio recordings. Where applicable, refer to IOM 5.3.5 for procedures on how to prepare your own recording in parallel with the firm's recording.

It is the Agency's position that while the investigator must protect privileged information provided to him/her during the inspection, it is the firm's responsibility to protect privileged/confidential information observed or recorded by those individuals invited by the firm.

5.1.5 - General Procedures & Techniques

The procedures and techniques applicable to specific inspections and investigations for foods, drugs, devices, cosmetics, radiological health, or other FDA operations are found in part in the IOM (inspectional and investigational policy/procedure), various Guides to Inspections of... (a "how to" guidance series), and the Compliance Program Guidance Manual (program specific instructions). Some procedures and techniques which may be applicable to overlapping areas or operations are as follows:

5.1.5.1 - Candling

Candling is defined as: "to examine by holding between the eye and a light, especially to test eggs in this way for staleness, blood clots, fertility and growth." Like most techniques learned through the food inspection programs, there are uses for this technique in other program areas such as looking for mold in bottled liquids which could be drugs, devices or biologics. Candling can also be useful in the examination of original documents to see below white-out or to look for over-writing.

Many types of products lend themselves to inspection by some type of candling. For these products, firms generally have candling equipment which may be built into the production lines or may be a separate operation.

Where checking products by candling, it may be possible to utilize the firm's candling equipment. Various other light sources for candling are also available including overhead projectors. Exercise care when using overhead projectors and protect the glass surface and the lens from scratches and damage. All candling is best accomplished when light outside the item being candled is masked so the light passes through the object rather than being diffused around it. A heavy paper or cardboard template can be quickly prepared at the time candling is done.

5.1.5.2 - Label Review

Do not undertake a critical review of labels unless instructed by the assignment, program, or your supervisor. Limit your comments to the mandatory label requirements required by the Acts. However, if after review of the formula, it is obvious an active ingredient or an otherwise mandatory ingredient statement does not appear on the label, such discrepancy may be called to management's attention. See also IOM 5.2.3.2 regarding labeling for blood and blood products.

If asked for other label comments, refer the firm to the appropriate Center to obtain a label review.

When the labeling is suspect or when you are requested to collect labels/labeling, collect three copies of all labels and accompanying literature for further review. For medical devices, if there is a question regarding the need for a new 510(k) or PMA supplement, it is essential the label and labeling be collected.

5.1.5.3 - Field Exams

A field examination is an on-site examination of a domestic product (or a foreign product in domestic channels of trade) sufficient in itself to determine if the product is in compliance with the Acts enforced by FDA.. A field exam can be conducted of any commodity in any location. If the examination does not reveal a violation or the appearance of a violation, a sample of the lot is usually not collected. If your exam reveals a violation or potential violation, you should collect an official sample. With the implementation of FACTS, your time spent conducting the field exam is reported even if you do collect a sample. Only the actual time spent in the collection of the sample would be reported as sample time.

Instructions on how to conduct a field exam are contained in "Guides To The Inspection of ***" and Compliance

Programs. The Sample Schedules in Chapter 4 also
provide guidance on lot examinations for special situations.

5.2 - Inspection Procedures

5.2.1 - Pre-Inspectional Activities

Prior to the start of any inspection or investigation, you
should conduct a number of activities. These will differ
based on whether this is an inspection or an investigation.
You should a review of the establishment's factory jacket (if
one exists), and registration and listing (if applicable)
information. The purpose of this review is to determine the
location of the establishment and obtain an overview of the
establishment's operations and products as well as an
understanding of their compliance history. You should also
evaluate the establishment factory jacket to determine if
there were any prior safety issues noted, e.g. documented
Investigator safety incidents or whether any specific
personal protective equipment is needed prior to the start of
the inspection. If there has been a past personal safety
incident, you should discuss with your supervisor and
develop a Situational Plan prior to the start of the
inspection. See IOM 5.2.1.4 – Personal Safety Plan.

Prior to initiating any inspection you should become familiar
with the reporting requirements for the specific assignment,
as well as the requirements of IOM Subchapter 5.10.

If the inspection or investigation is a directed assignment
from a Center, ORA headquarters or another district, read it
and attached materials to assure you understand the
assignment. If the inspection or investigation is being
conducted in part or solely as a recall follow-up or
complaint, refer to Chapter 7 (Recalls) or Chapter 8
(Investigations) of the IOM for additional guidance.

You should review the applicable FACTS assignment to
determine if the Personal Safety Alert indicator is checked
for this specific firm. The reason for the Personal Safety
Alert should be listed in Endorsement and should be
accompanied by a Memo to the Establishment File Jacket
or documented in a prior EIR. See IOM 5.2.1.3 Personal
Safety Alert.

You should also review the applicable Compliance Program
Guidance Manual(s) prior to the start of your inspection or
investigation. ORA's Division of Field Investigations (DFI)

has written numerous Inspection Guides to assist you in conducting inspections of various types of establishments, products or processes. You should become familiar with the appropriate guides prior to the start of the inspection and utilize them as needed throughout the inspection. The Centers have issued numerous guidance documents for industry. These documents are normally posted to the appropriate Center's Internet and Intranet web sites.

Subchapters 5.4-5.9 of the IOM contain additional, program specific pre-inspectional activities, which you should follow.

Imported products cross all program areas and our regulation of them does not stop at the border. Determine if there are any "import for export" follow-up assignments and be prepared to cover them during your inspection. See IOM 6.2.3.4 for guidance. Please be alert to imported products whenever you make an inspection. During inspections of domestic firms, if you encounter imported products that appear adulterated, misbranded, counterfeit, tampered with or otherwise suspect, attempt to fully identify the product and the source of the imported products. Contact your supervisor and DIOP (HFC-170) if necessary.

5.2.1.1 - Pre-Announcements

Pre-announcements are mandatory for all medical device inspections in accordance with the criteria and instructions below and BIMO sponsor/monitor inspections. In all other program areas, pre-announcements may be made at the discretion of the district. If you are going to visit facilities where livestock (including poultry) or wild animals are housed or processed, review IOM 5.2.10. In general, it may be inappropriate to pre-announce inspections of food establishments, blood banks, source plasma establishments and some BIMO inspections, but this too is subject to district discretion. If a district believes pre-announcing an inspection of an establishment will facilitate the inspection process then the procedures below for doing pre-announcements for medical device inspections should be followed. ORA's primary purpose for pre-announcing is to assure the appropriate records and personnel will be available during the inspection. It is not to make an appointment for the inspection. It should not be referred to as an appointment to inspect. When doing a pre-announcement, it is important you communicate to the establishment the purpose of the inspection and a general idea of the records you may wish to review. If you find

Investigations Operations Manual Chapter 5:
Establishment Inspections

neither the appropriate personnel nor records available, note this in your Establishment Inspection Report (EIR). The District may use this data in the future when considering whether this establishment should be eligible for pre-announced inspections.

The following is the general outline for pre-announcement of medical device inspections. You are advising the establishment's management of the date and time you will be arriving at the establishment to conduct the inspection. The establishment has no authority to negotiate this. If you, as the investigator, feel the need to accommodate the establishment's request, be sure there are sound reasons for doing so and report them in your inspection report.

5.2.1.1.1 - Basic Premises

Pre-announcement of inspections is to be applied only to establishments that meet specific criteria. Pre-announcement may be considered for establishments that manufacture both drugs and devices or biologics and devices. The eligibility of an individual establishment for pre-announced inspection is at the discretion of the inspecting office using clearly described criteria. (See Criteria for Consideration) The district does not have the discretion to decide the types of medical device establishments eligible for pre-announcement, but may decide the specific establishments' eligibility because they meet the criteria.

The pre-announcement should generally be no less than 5 calendar days in advance of the inspection. Should a postponement be necessary, the decision as to rescheduling rests with the investigator/team, but the new inspection date should not be later than 5 calendar days from the original date. Inspections may be conducted sooner than 5 calendar days if requested by or acceptable to the establishment and if this date is acceptable to the investigator/team.

To participate in the pre-announcement portion of the program, establishments are expected to meet the commitment to have appropriate records and personnel available during the inspection.

Pre-announced inspections will not limit an investigator's authority to conduct the inspection. Inspections will be as thorough as necessary.

5.2.1.1.2 - Criteria for Consideration

When deciding whether an establishment qualifies for a pre-announced inspection, you must consider whether both the type of inspection and the establishment's status meet the following specific criteria.

5.2.1.1.2.1 - Type of Inspection

Only the following types of inspections are appropriate:

1. Pre-market inspections (PMA, 510(k))

2. Foreign inspections

3. Quality System/Good Manufacturing Practice (QS/GMP) inspections:
    a. Biennial routine inspections
    b. Initial inspections of new facilities or newly registered establishments
    c. Initial inspections under new management and/or ownership.

5.2.1.1.2.2 - Eligibility Criteria

Establishment's eligible for pre-notification should meet the following requirements:

1. Non-violative QS/GMP inspection histories (inspections classified as no action indicated (NAI) or voluntary action indicated (VAI)). For VAI, adequate corrections of conditions observed and listed on FDA 483 during the previous inspection were verified and did not lead to any further agency action.

2. To remain eligible for pre-announced inspections, establishments must have a history of having individuals and/or documents identified in previous pre-announced inspections reasonably available at the time of the inspection.

5.2.1.1.3 - Procedures

Procedures:

1. The investigator designated to conduct the inspection will contact the most responsible individual at the facility. You should leave a message requesting a return call if the most responsible person at the facility is unavailable at the time the call is made. The district should use good judgment as to what is a reasonable time frame to await the return call.

2. Changes in dates should be kept to a minimum. If a change is made, a new date should be provided as soon as possible, which will facilitate the inspection and accommodate the investigator's schedule. The establishment should provide a valid reason for requesting a change in the start date. A valid reason should be the same as you would accept if presented with the information during an unannounced inspection.

3. Inform the establishment as to the purpose, estimated duration, and the number of agency personnel expected to take part in the inspection. The products or processes to be covered should be described if this will facilitate and be consistent with the objectives of the inspection.

4. When known, specific records/personnel will be requested at the time the inspection is pre-announced.

5. The notification should be as specific as reasonably possible and specify the date for the start of the inspection.

Include in your EIR whether or not the inspection was pre-announced and include information on any difficulties experienced in notification or accessing records or personnel, which should have been available as a result of pre-announcing the inspection. For medical device establishment inspections, if not pre-announced, describe briefly in the EIR why not. If an establishment should become ineligible for pre-announcement, the endorsement of the EIR should include this statement. This information will be necessary for making a determination regarding future pre-announced inspections of the establishment. In addition, it is advisable to inform the establishment during the current and subsequent inspections of the action(s), which may have caused them to be ineligible for pre-announcement.

Subchapters 5.4-5.9 of the IOM contain additional, program specific pre-inspectional activities, which you should follow.

5.2.1.2 - Personal Safety

ORA considers the safety of Investigators, Inspectors and all those who meet with regulated industry to be of the utmost importance. Personal safety concerns are defined as those factors FDA employees should maintain awareness of which potentially affect their safety during an inspection, such a threatening situation; or where specific personal protective safety equipment is warranted; or where a particular inspection may be medically contraindicated for specific FDA personnel. When these conditions are noted during an inspection, the Investigator should discuss the situation with their Supervisor and ensure that the Personal Safety Alert is checked in FACTS and a Memo to the File is generated– see IOM 5.2.1.3 For information concerning personal protective equipment, see IOM Subchapter 1.5.

Physical resistance to FDA inspections and threats to, or assaults on, FDA employees engaged in their work are extremely rare. However, there will be times you are confronted by unfriendly or hostile persons. ORA has offered various conflict resolution training courses to assist and prepare you for how to diffuse a situation. In most instances, conducting your activities with tact, honesty, diplomacy, and persuasiveness will be enough to diffuse the situation. While at times, you may have to adopt a firm posture, you should not resort to threats, intimidation, or strong-arm tactics. Refer to IOM 5.2.5.4 for Hostile and Uncooperative Interviewees.

Safety is the responsibility of all FDA employees, including you, your supervisor and other Agency management. When you receive an assignment, it is important to evaluate the assignment not only in accordance with IOM Section 5.2.1, but also with respect to your personal safety. If you determine there is the possibility of a threat to your personal safety, consult with your supervisor. You and your supervisor should consider developing a Situational Plan in preparation for the inspection.

5.2.1.2.1 - Preparation

Below are some suggested items the District may consider when preparing for your next assignment to assess if there are potential personal safety issues. This list is not meant to be all inclusive.

1. Does the assignment involve working with other Federal Agencies such as U.S. Marshals, Federal Bureau of Investigations, US Customs in executing search warrants, seizures etc?

2. Does the assignment involve working with or contacting FDA's Office of Criminal Investigation?

3. Does the assignment involve a firm where there is a suspicion and/or knowledge of questionable or illegal activities?

4. Does the assignment involve a suspected tampering and/or a visit to an individual's residence?

5. What is the past history from a personal safety standpoint with the prior interactions with representatives of this firm? Have the FDA's State counterparts or other Federal and/or local agencies indicated a concern for personal safety? What does the firm's establishment file indicate about personal safety over the past inspections?

6. What is the location of the firm or the operation? Is it in an area which may be unsafe? Have the inspected firm or any of it's employees been uncooperative with government officials?

7. Is the firm known to the Agency? Has the Agency any additional information which would assist in your evaluation?

If these questions and/or others result in a concern for your personal safety, then a Personal Safety Plan should be developed and approved by District management before conducting the assignment. See IOM 5.2.1.4 - Situational Plan.

Due to the unlimited variability of potential safety situations, it is not feasible to prescribe in the IOM what to do in every instance. The decision of what to do in each individual circumstance rests with the Investigator and their District management. Your district management is most familiar with the specific firm in question, the regulated industry, as well as other local Federal, State and Local officials who may be able to provide you additional information and assistance. In addition, the experience of your District management combined with the various training courses on conflict resolution may also be consulted. Districts should notify Division of Field Investigations to inform headquarters of any potential safety concern, so that DFI may track personal safety issues. DFI will also maintain a library of Personal Safety Plans which may also be of use to your District. DFI may be contacted at 301-827-5653 or at the following personal safety e-mail address: ORAHQDFICSOSAF@FDA.HHS.GOV

5.2.1.2.2 - Physical Resistance/Threats/Assaults

If you receive physical resistance or threats, or if you sense the real possibility of an assault, disengage from the confrontation, get to safety, and call your supervisor immediately. Make careful and exact notes later of who said what to whom, who did what, and whether someone tried or succeeded in threatening, assaulting or taking information or equipment or samples from you. Be careful in any descriptions you give or write of such events, just as you are in recording other evidence that may result in a court case. Your safety is more important to the United States than the inspection or the sample. FDA will work with law enforcement government officials, e.g. FDA's Office of Criminal Investigations' (OCI) Special Agents, local police, or United States Marshals to assist an inspection team if there is a reasonable fear of danger to the investigator.

If you are assaulted (either physically or put in fear by threats of physical violence), your supervisor can summon local police, United States Marshals, FBI or contact OCI headquarters for assistance (301-294-4030). While OCI does not normally provide physical security in these cases, they will assist in threat evaluation based on specific facts and available criminal databases. OCI can also make contacts with local police and federal agencies based on previous established liaisons. If you have been assaulted or threatened and you are unable to reach your supervisor or other District management, you should contact the local police in the area where the assault or threat occurred. Be careful in any descriptions you give or write of such events, just as you are in recording other evidence that may result in a court case. Make sure that any inspected facility where weapons are observed, or where threats or assaults occur, is identified on that facility's Endorsement page of the inspection report for that facility and to your supervisor, so that Investigators or Agents who follow you into that facility will be alert to those possibilities. Your supervisor would also be responsible for checking the Personal Safety Alert box in FACTS and for beginning the notification process to alert other Federal or State agencies that also inspect the facility of the possible danger. For more information see IOM 5.2.1.3 Personal Safety Alert. For specific safety guidance related to inspections and interviews, see IOM 5.2.5.4.2 Hostile and Uncooperative Interviewees.

In addition, in any instance where you have perceived a threat to your personal safety during an inspection, investigation or sample collection, you should exit the situation immediately and report it to your supervisor. You should then write a memorandum of the event in a factual manner including information pertaining to the who, what, when, where, and how of the event. Be careful in any descriptions you give or write of such events, just as you are in recording other evidence that may result in a court case. This memo will be filed in the official establishment file jacket and copies be sent to any and all resident posts and import offices who may interact with this firm. The memo will be filed on the opposite side of the folder from all other documents and will be a printed on eye-catching color paper in order for the document to be visible to the next Investigator. The memo should be retained and maintained within the District. A copy of the Memo documenting the personal safety situation should also be sent to Division of Field Investigations, HFC-130.

5.2.1.3 - FACTS Personal Safety Alert

Within the Maintain Firms Option in the FACTS system, there is Personal Safety Alert option that allows the Supervisor (FACTS Supervisor Role) to check the appropriate box to advise the FDA Investigator that there is a personal safety issue. Only the FACTS Supervisor Role will allow for updating the Maintain Firms screen. This personal safety alert may be selected when there is a potential hazard identified:

1. Where specific personal protective equipment is needed (respirators etc)

2. Where a previous threat/assault or physical resistance occurred

3. Where there are specific medical considerations for a population of Investigators (e.g. the firm manufactures a drug hazardous to women of child-bearing years or those with allergies to peanuts, penicillin, or other products.)

In any example listed where there is a Personal Safety Alert, the specific safety alert should be documented both in the Endorsement and in a Memo to the File. The memo should be flagged "MEMO TO FILE — PERSONAL SAFETY ALERT" and should provide the factual information to support why the Investigator should be alerted to the safety issue. Be careful in any descriptions you give or write of such events, just as you are in recording other factual evidence that may result in a court case. The memo should be filed in the official establishment file jacket and copies sent to any and all Resident Posts and Import Offices who may interact with the firm. The memo will be filed on the opposite side of the folder from all other documents and will be a printed on eye-catching color paper in order for the document to be visible to the next Investigator. The memo should be retained and maintained at the District Office. A copy of the Memo documenting the personal safety situation should also be sent to Division of Field Investigations, HFC-130. The Supervisor and/or other District management will be responsible for evaluating any corrective actions taken by the firm or individual to remove or stop the potentially dangerous situation or condition. If the situation remains potentially dangerous, the Personal Safety Alert should be maintained in FACTS. Follow-up inspections at the facility should continue to document whether or not the safety situation continues exists. If the situation has been resolved (new management, dismissal of an employee, cessation of penicillin in a facility, etc), the Personal Safety Alert should be removed from FACTS by the supervisor.

5.2.1.4 - Personal Safety

A Personal Safety plan is an investigative tool developed to assist in managing and preparing for a potentially dangerous situation. Districts should consider developing a Personal Safety Plan when the conditions surrounding the specific inspection, investigation or sample collection indicate a plan is needed. The plan allows all those involved to carefully evaluate the specific inspection in order to prepare for a successful conclusion. Utilizing Personal Safety Planning concepts prior to a potentially dangerous situation is supported by the Federal Law Enforcement Training Center and is part of the training programs of many other Federal Agencies. The plan should document what specific roles and responsibilities are needed to conduct the inspection/investigation or sample collection. The plan should also answer the questions: Who, What, Why, When and Where concerning the potential danger.

There are seven principles to a Personal Safety Plan. These are:

1. *Summary of Potential Hazards:* This section of the personal safety plan includes all of the potential hazards, in a detailed description, that prompted the need for a personal safety plan. Be sure to answer the questions: Who, What, Where, When, and Why. Also include any specific hazards that require personal protective equipment or situations at the facility that may cause allergic reactions for CSOs/Analyst. Include in the section information from past inspection reports, discussions with past FDS, State or local investigators, as well as any environmental or plant/facility specific information that would negatively impact a successful personal safety plan when initiated.

2. *Sources of Information:* This section of the personal safety plan includes all the sources from where your potential hazards were collected. For instance, document which CSO or State inspector supplied their factual statements; state which documents/databases the information was from that assisted in your hazard summary. This section is important, as it documents factual evidence, similar to all of you other FDA factual inspection gathering information.

3. *Response Alternatives:* In this section, provide a list of factual, practical responses or options to consider. This will also allow your supervisor to see all the possible ways to handle the situation.

4. *Response Plan:* This section will be the most important part of your plan because it includes all of the details of what will be done to mitigate the hazards

A. The response plan should also include all of the tools that your possess to assist you in handling the situation carefully. This should include training, experience, and other tools/procedures you have at your disposal. Roles and Responsibilities of all involved in the plan will be identified include those intended to be on-site, and those who will be off-site, and participating in the plan.

B. *Communication:* provide all information about how communication will occur between on/off site; amongst those present on-site, and any emergency, law enforcement or medical responders. Also consider types of communication code words for emergency.

C. *Transportation:* Provide information in the plan as to how travel to the facility will happen. Is there a coordination point? Do you intend to use Government marked or unmarked cars? Who will ride in each car. What route will be taken going to and leaving the facility. Consider where you will park the car when you arrive at the facility. Consider what modes of communication will be used to communicate if multiple vehicles are used.

D. *Equipment:* Include in this section all equipment needed to initiate this plan. Is personal protective equipment needed? Is there any special sampling equipment or other equipment needed? Include in this section, equipment such as communication tools, FDA forms, etc. Assure that the equipment needed is in full functioning mode.

E. *Emergency Exit Strategy:* Describe in this section what the exit strategy will be in the event of an emergency. Consider emergency strategies for safety (issues), as well as any medical emergency. How will the emergency be communicated on-site and off-site? How do you exit the facility and return to your vehicle? Is there a scheduled meeting point to assure all are safe? Remember, we do not want to leave any one behind.

Remember to contact your supervisor when you return to safety.

5. *Evaluation:* Once the plan has been completed, a debriefing of the situation should occur with all who

were involved in the plan development. Evaluate what went well, what needed improvement, what would we do next time. Evaluate whether the plan was successful and document lessons learned for the next time.

The Personal Safety Plan should be developed by the Investigator, Supervisor, other CSO's who may be familiar with the facility, Compliance Officer, if needed, and any other individuals (District, Region, or HQ experts, etc.) who may be able to assist in the depth, scope, and specifics of the firm in question. The decision of who should be involved in the development and approval of the plan is left to the Districts' discretion.

District management and all involved in writing the Personal Safety plan should meet when necessary in order to assure a well developed, and understood personal safety plan. You and your supervisor should maintain contact during the execution of the personal safety plan. The Supervisor should contact the employee during these personal safety situations at a predetermined frequency outlined in your plan. A debriefing session should be held following the execution of the plan. Discussions should include what actions worked well and where there are areas of improvement.

For foreign inspections where a Personal Safety Plan is warranted, DFI will assist the inspection team. The inspection team's management may also wish to participate so that there is clear understanding of what actions will be taken for the foreign inspection.

The Personal Safety Plan should be placed in the official establishment file jacket separately from any EIRs in the same location as any Personal Safety Alert memos. A copy of completed and executed Personal Safety Plans must be sent to DFI (HFC-130) in order for DFI to maintain a reference library of all Personal Safety Plans.

5.2.2 - Notice of Inspection

Upon arrival at the firm locate the owner, operator or agent in charge of the establishment. This should be the top Management Official on site. Be certain of this individual's status. Introduce yourself by name, title and organization. Show your credentials to this person and present a properly signed, completed, original of the FDA 482, Notice of Inspection.

If additional Agency personnel accompany you during the inspection, they must show their credentials to the top Management Official upon arrival at the site. A new FDA 482, Notice of Inspection must be issued. Submit the carbon copy of the FDA 482(s) with your EIR. Explain the purpose of your visit. Readily accept any management offer to have a representative accompany you on the inspection.

If non-FDA officials accompany you during your inspection and do not have authority to enter and inspect, you should obtain permission (preferably in advance) from the most responsible individual at the firm. Non-FDA officials and those who do not hold FDA credentials do not sign the FDA 482. See IOM 5.1.1 and 5.10.4.3.3.

For multiple occupancy inspections in drug establishments, refer to IOM 5.1.1.11. Inspections of multiple firms, which are separate legal entities, should be reported under separate EIRs.

If faced with a refusal, or partial refusal of inspection proceed as outlined in IOM 5.2.5.4.

Any time a FDA 482 is issued, also issue FDA 484, Receipt for Samples, if you collect any samples at the firm. See IOM 5.2.4. See IOM 4.1.1.1 and 4.1.1.2 for instructions for issuance of the FDA 482 in certain sampling situations.

See IOM 4.2.4.3 and 4.2.4.4 for situations where you would issue an amended FDA 482 for sample collections only. The FDA 482 may be amended "To Collect Samples Only" as shown in IOM Exhibit 5-4.

If you have concerns of when to or when not to issue the FDA 482, discuss with your supervisor.

5.2.2.1 - Multiple Date Inspections

If your inspection covers more than one day, advise management at the close of each day you have not finished the inspection and when you will return. Do this each day until you finish the inspection. A FDA 482 is not required for each day of an inspection or when different individuals are interviewed. If there will be an extended period of time (i.e., a week or longer) before you can return to the firm to complete the inspection, be sure management is aware of the delay and discuss with your supervisor whether or not you need to issue another FDA 482.

5.2.2.2 - Inspection of Vehicles

If vehicles are present which are owned or leased by the firm being inspected and it is necessary to inspect the vehicles, the inspection of these is covered by the FDA 482, Notice of Inspection, you issued to the firm.

If vehicles (trucks, trailers, RR cars, etc.) which are not owned or leased by the firm are present and inspection is necessary, a separate FDA 482, Notice of Inspection, is required:

1. Issue the FDA 482 to the driver of the vehicle.

2. If the driver is not present and if, after a diligent search, he cannot be located, issue a separate FDA 482 jointly to the firm being inspected and to the firm whose name appears on the cab. Enter the license number of the vehicle on the FDA 482. Give the original FDA 482 to the firm and leave a copy in the cab of the vehicle.

3. If there is no cab present, prepare a separate FDA 482 modified to read "*** to inspect unattended vehicle ***" and issue it to the firm being inspected as the "agent in charge" of the vehicle. Enter the license number of the vehicle, trailer or RR car number, etc., on the FDA 482. Should the firm being inspected refuse to accept the Notice, leave it in a conspicuous place in the vehicle. Describe the circumstances in your EIR.

5.2.2.3 - Follow-Up Inspections by Court Order

At times you may be instructed to conduct inspections of firms by authority of an injunction or other court order. This situation provides separate and distinct inspectional authority involving both the authority of the court order and the authority of Section 704 of the FD&C Act [21 U.S.C. 374], each providing independent courses of action.

When assigned to conduct inspections under these situations, obtain a copy of the injunction or other court order bearing the filing stamp and all relevant signatures. Prior to starting the inspection study the order thoroughly for any special instructions of the court. Your supervisor will assist you in determining the depth of the inspection necessary to cover all of the court requirements.

Take a clearly legible copy of the court decree (not necessarily a certified copy) with you to the firm to be inspected.

Present your credentials in the same manner as for any other EI. Issue the FDA 482, Notice of Inspection, modified to read, "Notice of Inspection is hereby given under authority of injunction (provide here the injunction number and/or other identification) against the firm and pursuant to Section 704 ***". Show the person to whom the FDA 482 was issued a copy of the Order, and, read the following statement to that person.

"This inspection is being conducted under the authority of injunction (add the injunction number and/or other identification) (or other court order) granted by the United States District Court against this firm on (date). The inspection will cover all items specified in the decree. In addition to the inspection authority granted in the court decree, I am issuing you a Notice of Inspection under the authority of Section 704 of the Federal Food, Drug and Cosmetic Act which authorizes inspections of firms subject to that Act."

If, the firm refuses access to records, facilities, or information for which the decree provides inspectional authority, read the pertinent section(s) or portion of the order to the person refusing so there will be no misunderstanding as to the requirements of the decree. If the person still refuses, report the facts to your supervisor as soon as possible so the court can be promptly advised of the situation. See IOM 5.2.5 for information on handling refusals.

At the conclusion of the inspection and a FDA 483 is to be issued and you are using Turbo EIR, follow the Turbo instructions to get injunction specific cites on the FDA 483.

When you prepare your EIR, describe the sequence of events in detail including exactly what happened and how you handled the situation. This documentation will help support any charge of violating the court order and/or Section 704 of the FD&C Act [21 U.S.C. 374].

The court order may require a report to the court. Discuss this with your supervisor since the district will normally handle this part of the requirement.

5.2.2.4 - Conducting Regulatory Inspections When the Agency is Contemplating Taking, or is Taking, Criminal Action

You should not issue a Notice of Inspection if the agency is contemplating taking, or is taking, criminal action against a firm without first discussing the matter with your Supervisory Investigator. Federal Rules of Evidence may not permit using evidence in a criminal matter if it is knowingly obtained under administrative authorities such as Section 704 of the FD&C Act [21 U.S.C. 374]. It is the responsibility of the office generating the inspection assignment to inform the District if a criminal action is ongoing or contemplated. Once alerted, the Supervisory Investigator will then obtain advice from the Office of Chief Counsel and, once obtained, will assign the inspection to the Investigator(s).

Decisions to inspect under such circumstances should be based on considerations of whether or not the request is consistent with FDA's responsibility to assure articles are not produced or distributed in violation of the Federal Food, Drug, and Cosmetic Act or other Federal law within FDA's jurisdiction. It would be lawful to conduct an inspection to identify such violative products and to determine if corrective action was necessary to bring such products into compliance. However, it would be an abuse of the regulatory inspection authority for FDA to conduct a regulatory inspection under that authority for the sole purpose of gathering evidence of criminal violations. Such an abuse is unlawful, and could have significant consequences.

This is because, in general, the Fourth Amendment to the United States Constitution prohibits searches without a warrant. One exception to the warrant requirement includes the inspection of industries long subject to close supervision and inspection, which are conducted under a statute that dispenses with the need for a warrant. Because such inspections are not subject to advance scrutiny for probable cause, as would be an inspection conducted pursuant to a criminal warrant, the Supreme Court has warned government entities not to use administrative inspections to search for criminal violations in an effort to sidestep the Fourth Amendment. So long as the Agency conducts the administrative inspection in good faith for a valid, non-criminal purpose, evidence gathered in such inspections generally may be used in a criminal prosecution. However, the facts of each case are unique, and employees involved must carefully document the Agency's purpose in conducting the inspection.

Because the Agency's underlying purpose in conducting an inspection ultimately will determine whether the inspection was conducted in good faith to pursue a valid, non-criminal purpose it is important to document the non-criminal purpose for an inspection undertaken under these circumstances. The need for and extent of such documentation is at a minimum when the non-criminal purpose of the inspection is evident and compelling, for example, when the purpose is to determine articles are being produced in conformity with the Food, Drug, and Cosmetic Act. The need to document the non-criminal purpose of the regulatory inspection increases as the likelihood of criminal prosecution increases. For example, there would be an increased need to document the regulatory purpose of an inspection if the matter has been referred to the Department of Justice for grand jury investigation.

There may be occasions when neither the office generating the inspection assignment nor the District conducting the inspection is aware the Office of Criminal Investigations is conducting a criminal investigation of a firm that is the subject of a regulatory inspection. The Office of Criminal Investigations may determine it is not in the interest of the agency to disclose to other components of FDA the existence of its investigation, as long as the Office of Criminal Investigations is not involved in the agency decision to conduct a regulatory inspection. However, the Office of Criminal Investigations and other components of FDA may also share information as set out below.

5.2.2.5 - When Evidence of a Criminal Violation is Discovered in the Course of a Regulatory Inspection

There may also be occasions where you are conducting a regulatory inspection at a facility, and, in the course of that inspection, you discover evidence of a criminal violation. If this occurs, you should continue the regulatory inspection as you would under normal circumstances. Document the observation and notify your supervisor. Evidence of the observation could be used in a criminal investigation, and the evidence could legally be disclosed to criminal investigators.

If you know criminal investigators are conducting a criminal investigation, your supervisor should notify the criminal investigators of any such observations. If you do not know of any ongoing criminal investigation, your supervisor should refer the information for review by the Office of Criminal Investigations. See the current Regulatory Procedures Manual (RPM). If the regulatory inspection is Center-directed (such as a bio-research monitoring inspection, a pre-approval inspection, or an inspection related to data integrity issues) your supervisor should immediately notify the Center involved of the referral to the Office of Criminal Investigations.

The discovery of evidence of a criminal violation may also be relevant to FDA's responsibility to assure articles are being produced in conformity with the Food, Drug, and Cosmetic Act. Additional inspections may be warranted. Such inspections should be planned and documented in accordance with the preceding section, "Conducting Regulatory Inspections When the Agency is Contemplating Taking, or is Taking, Criminal Action."

5.2.2.6 - Use of Evidence Gathered in the Course of a Criminal Investigation

The extent to which information gathered in the course of a criminal investigation may be shared with other components of FDA will vary with each case. Investigators should determine the extent of information sharing in accordance with the following guidelines.

Information and evidence gathered in the course of a criminal investigation may be shared with regulatory personnel, subject to two reservations:

1. Information obtained pursuant to grand jury subpoena or testimony may not be shared. Disclosure of such information to anyone other than individuals identified by the Department of Justice attorney involved could subject the individual making the improper disclosure to sanctions for contempt by the court. Only the court can authorize disclosure beyond these parameters. Information obtained by other means (search warrant, cooperative witnesses, surveillance, etc.) may be shared, subject to the following paragraph.

2. There may be a need to protect the confidentiality of the criminal investigation. For example, disclosure to

regulatory investigators might prematurely disclose the existence of the criminal investigation or the identity of confidential informants. However, whenever you are calculating the need to protect the confidentiality of information gathered in the course of a criminal investigation through means other than the grand jury, you must consider whether it will be in the interest of public health to protect the confidentiality of that information.

Criminal investigators should consult their supervisors to determine whether disclosure should be made to regulatory investigators.

5.2.2.7 - Use of Evidence Voluntarily Provided to the Agency

Criminal and regulatory investigators may share information and evidence voluntarily provided to FDA, without use of the regulatory inspection authority, search warrant, or subpoena. If criminal investigators decide not to share such information because of a need to protect the confidentiality of the criminal investigation, they should consider the potential impact on the public health of protecting the confidentiality of that information.

5.2.2.8 - Concurrent Administrative, Civil, and Criminal Actions

It may be appropriate to seek administrative and/or civil remedies against a firm or individual under investigation for criminal violations. There are many issues involved in determining whether such actions may proceed concurrently, or whether certain actions should proceed first. Each situation must be evaluated on an individual basis. If administrative and/or civil remedies are under consideration against a firm or individual also under investigation for criminal violations, representatives from the Center responsible for evaluating the administrative and/or regulatory action should meet with the Office of Criminal Investigations Headquarters staff to issues related to the timing of administrative, civil, and criminal actions. The Office of Criminal Investigations and other components of FDA may share information subject to the reservations set out earlier.

5.2.2.9 - Working with a Grand Jury

Finally, if you are assigned to work with a grand jury, you should not participate in a regulatory inspection or other regulatory matter involving the same firm or individual(s). Such participation is contrary to long standing agency policy, might be unlawful, and could result in sanctions against the investigator and the agency. You should not participate in any regulatory matters that could result in improper disclosure of grand jury information, even after the grand jury investigation is closed. Grand jury proceedings remain secret even after they are concluded. Under no circumstances should you undertake such participation without first obtaining clearance from the Department of Justice attorney or the Office of Chief Counsel attorney assigned to the grand jury case. See IOM 2.2.7.3 for additional information on Grand Jury proceedings.

5.2.3 - Reports of Observations

The FORM FDA 483 INSPECTIONAL OBSERVATIONS (see Exhibit 5-5) is intended for use in notifying the inspected establishment's top management in writing of significant objectionable conditions, relating to products and/or processes, or other violations of the FD&C Act and related Acts (see IOM 5.2.3.2) which were observed during the inspection. These observations are made when in the Investigator's "judgment" conditions or practices observed, indicate that any food, drug, device, or cosmetic have been adulterated or are being prepared, packed, or held under conditions whereby they may become adulterated or rendered injurious to health. The issuance of written inspectional observations is mandated by law and ORA policy.

Be alert for specific guidance in assignments or Compliance Programs which may supplement the following general instructions.

All FDA-483s should adhere to the following general principles:

1. Observations which are listed should be significant and correlate to regulated products or processes being inspected.

2. Observations of questionable significance should not be listed on the FDA-483, but will be discussed with the

firm's management so that they understand how uncorrected problems could become a violation. This discussion will be detailed in the EIR.

All FDA-483s should have the following characteristics to be useful and credible documents:

1. Each observation should be clear and specific.

2. Each should be significant. Length is not necessarily synonymous with significance.

3. Observations should not be repetitious.

4. The observations should be ranked in order of significance.

5. All copies of the FDA-483 should be legible.

If an observation made during a prior inspection has not been corrected or is a recurring observation, it is appropriate to note this on the FDA 483.

As of 1997, ORA established a FDA 483 annotation policy for medical device inspections. See IOM 5.2.3.4. Regardless of whether an establishment's FDA 483 is annotated, investigators and analysts should make every reasonable effort to discuss all observations with the management of the establishment as they are observed, or on a daily basis, to minimize surprises, errors, and misunderstandings when the FDA 483 is issued. This discussion should include those observations, which may be written on the FDA 483 and those that will only be discussed with management during the closeout meeting. Industry may use this opportunity to ask questions about the observations, request clarification, and inform the inspection team what corrections have been or will be made during the inspection process. Investigators are encouraged to verify the establishment's completed corrective actions as long as the verification does not unreasonably extend the duration of the inspection.

There may be instances where same day discussion of observations may not be possible due to the volume of documents collected and document review reveals observations on a different day than the documents were collected or in other circumstances. When these instances occur immediately prior to the conclusion of the inspection the lack of a daily discussion of observations does not

preclude listing of significant observations which were not
previously discussed on the FDA 483.

Turbo EIR

Turbo EIR is an automated FDA 483 and EIR reporting
system. Use Turbo EIR to generate the FDA 483 where
applicable cite modules exist. Turbo EIR should not be
used to create a FDA 483 during an inspection of a firm
involving multiple commodity areas when FDA 483 cites do
not exist for ALL of the commodity areas for which
observations need to be included on the FDA 483. You
should be able to write the entire FDA 483 using Turbo EIR.

Use Turbo EIR for all EIRs whether or not your FDA 483
was generated using Turbo and when no FDA 483 was
issued. See IOM 5.10.4.

5.2.3.1 - Preparation Of Form FDA 483

It is not necessary to complete all headings of the FDA 483,
when multiple page 483s are issued. Complete all headings
on the first page and, on subsequent pages, only those
necessary to identify the firm and dates inspected. FDA
483s should be issued at the conclusion of the inspection
and prior to leaving the premises. However, in preparing
some complex FDA 483s, it may be necessary to leave the
premises and return at a later time to issue and discuss
your inspectional observations. In this case, you should
advise the firm's management your inspection has not been
completed and you will return to issue the FDA 483 and
discuss inspectional findings. There should be no
unreasonable or unwarranted delays in issuing and
discussing the FDA 483. During the inspection, do not show
the firm's management a draft, unsigned copy of the FDA
483 or an electronic copy of the FDA 483 on your computer
screen. You should issue only a signed FDA 483 at the
closeout discussion with management.

5.2.3.1.1 - Individual Headings

District Office address and phone number - Legibly print the
district address. Include the District Office commercial
telephone number and area code.

Insert this language in the address block of the form FDA
483 Inspectional Observations referencing the Information
for FDA - Regulated Industry website:

"Industry information: www.fda.gov/oc/industry".

1. Turbo EIR users - Turbo has been modified to automatically add the text and web address.

2. Adobe PDF or hardcopy FDA 483 forms – add or print the text and website address above below or beside the District Office Phone number.

See IOM 1.6.5.1 – Professional Stature for situations where firms express a concern during routine enforcement activities where an FDA 483 was not issued or the activity is not an inspection.

Name and Title of individual to whom report is issued - Enter legal first name, middle initial and last name and full title of the person to whom the form is issued.

Firm name - Enter full, legal name of the firm, including any abbreviations, quotation marks, dashes, commas, etc.

Street address, city, state and Zip Code - Enter street address, city, state and Zip Code. (Not P.O. Box unless P.O. Box is part of the address such as on a Rural Route).

Date(s) of inspection - Enter actual or inclusive date(s) of inspection.

FEI Number - If the FDA Establishment Identifier is on the assignment, enter it here. If not readily available, leave blank.

Type of establishment inspected - Enter the types of the establishment, such as bakery, cannery, wholesale warehouse, drug repackager, salvage warehouse, etc.

Employee(s) signature and Employee(s) name and title - The names of everyone who participated in the inspection with the issuance of a FDA 482 should be listed on the FDA 483 even if they are not available to sign the FDA 483. Each member of an inspection team should sign the FDA 483. However, absence of a team member at the conclusion of an inspection need not prevent issuance of the FDA 483. See IOM 5.1.2.5.1. If you use an electronically generated FDA 483, assure you have a copy for the District files -- an unsigned photocopy or printed duplicate is unacceptable. See IOM 5.2.3.6.2.

5.2.3.1.2 - Signature Policy

Everyone present at issuance signs the first and last pages of the FDA 483 and initials each intervening page in the signature block.

Note: *if you are not using the official multi-part FDA 483 form and a copier is not available, insert carbon paper to reproduce a signed copy of the FDA 483.*

See IOM 5.2.3.6 – Distribution of the FDA 483.

5.2.3.1.3 - Date issued

Enter date the form is actually issued to the firm's management.

5.2.3.1.4 - Observations

"During an inspection of your firm (I) (We) observed" - Where applicable, when formulating each FDA 483 observation, answer Who (using titles or initials when necessary), What, When, Where, How, and challenge each observation by asking So What? (regarding its significance)

Enter your reportable observations succinctly and clearly. Conditions listed should be significant and relate to an observed or potential problem with the facility, equipment, processes, controls, products, employee practices, or records. "Potential problems" should have a reasonable likelihood of occurring based upon observed conditions or events. Do not cite deviations from policy or guidance documents on your FDA 483.

As appropriate, FDA 483 observations should include relationship of observations to a given population, for example, "Two out of 50 records examined were * * *" or "4 out of 12 bags examined were ***." When appropriate, a FDA 483 observation may refer to inadequate situations as long as you provide supporting facts (examples) or explanation as to why the condition, practice or procedure observed is inadequate.

You should not identify individuals or firms by name on the FDA 483. Where appropriate to support the FDA 483 observation, identify the individual(s) or firm(s) by substituting other non-specific identifying information as below. Document your evidence in your EIR, fully explaining the relationship(s).

1. The lot number for a component received from or shipped to firm "A".

2. The invoice number for a shipment from or to firm "A".

3. A patient #, record #. See IOM 5.2.3.3 item 7.

4. The study number for a particular Clinical Investigator site.

5. Other necessary but non-specific identifying information to show the observation's relationship to a particular firm and/or individual.

FDA 483 Statements:

The following statement should be included on each FDA 483: "This document lists observations made by the FDA representative(s) during the inspection of your facility. They are inspectional observations, and do not represent a final Agency determination regarding your compliance. If you have an objection regarding an observation, or have implemented, or plan to implement, corrective action in response to an observation, you may discuss the objection or action with the FDA representative(s) during the inspection or submit this information to FDA at the address above. If you have any questions, please contact FDA at the phone number and address above."

Presently there are three ways to issue a FDA 483. Use the following guidance as it applies to the specific type of FDA 483 you are issuing.

1. Traditional hard copy FDA 483: This language is to be written or typed on the form until additional hardcopy forms are ordered. Note: it is only necessary to write or type this statement on the first page of a multi page FDA 483.

2. Electronic (non-turbo EIR) version of the FDA 483: Use the updated FDA 483 on the official forms web site to replace previous versions of the electronic version of the document in use.

3. Turbo EIR Field Agent incorporates this language in the FDA 483.

5.2.3.1.5 - Medical Device inspections

The following language should be inserted on the FDA 483 in addition to the above statement: "The observations noted in this form FDA 483 are not an exhaustive listing of objectionable conditions. Under the law, your firm is responsible for conducting internal self audits to identify and correct any and all violations of the quality system requirements."

5.2.3.1.6 - Correction of FDA 483 Errors

These procedures do not pertain to adverse conditions noted and then corrected during the inspection. Observations of this type stand and should remain on the FDA 483.

The Inspectional Observations (FDA 483) is of critical importance to both the Agency and regulated industry. Individual FDA 483s may become public through publishing in industry trade press, FOI inquiries, Headquarters postings and other means. Therefore, complete and accurate documentation of corrections to this official document is critical.

5.2.3.1.6.1 - Errors discovered prior to leaving the establishment

Non-Turbo, FDA 483s:

1. Make handwritten changes to correct the error/s on the original FDA 483 and initial the changes. Correct errors by striking through the erroneous text and entering the correct information (if any). When possible retrieve and destroy all uncorrected copies of the FDA 483 either provided to or produced by the establishment.

2. If the establishment has photocopying equipment available and will provide you with a copy of the corrected original FDA 483 then obtain a copy of the corrected original document from the establishment. If the establishment has no such equipment or refuses to provide you with you a copy of the original corrected FDA 483 then make the corrections and initial the changes using carbon paper and retain the carbon copy of the corrected FDA 483 for your District's official establishment file.

Turbo FDA 483s - All corrections/deletions should be made in Turbo.

1. Changes made to correct errors in the text of the observation will show on the face of the final printed FDA 483. Changed Text deletions will remain visible as strike through and correction made. For example, "lot 1234 5678" – (select text, right click, select font and select strike-through) or from "lot 1234" to "lots 1234 and 5678" and bold the changes "lots 1234 and 5678"

2. If an entire observation is removed, incidental text will be used to add the statement "An observation concerning *** was removed based on discussion with management."

3. Addition of a new item

5.2.3.1.6.2 - Errors discovered after leaving the establishment

Normally, you should not use addenda/amendments to issue additional FDA 483 items after the inspection has been closed out and you have left the premises.

1. Non-Turbo, FDA 483s (addenda): Discuss any errors with your supervisor. If necessary, an FDA 483 addendum limited to the corrected item(s) will be prepared.

2. Turbo FDA 483s (amendments): Discuss any errors with your supervisor. Make all corrections/deletions in Turbo. Changes made to correct errors in the text of the observation will show on the face of the final printed FDA 483. Changed Text deletions will remain visible as strike through and additions added.

3. Issuing FDA 483s (addenda/amendments): Personally deliver the FDA 483 addendum/amendment to the firm for discussion. If personal delivery is not practical, mail the addendum/amendment to the firm with a full explanation cover letter. Include a copy of the original FDA 483, FDA 483 addendum/amendment, and a copy of the letter in the EIR. In addition, you should call the person to whom the original FDA 483 was issued, to discuss the change(s). Document your discussion in your EIR.

5.2.3.2 - Reportable Observations

You should cite factual observations of significant deviations from the FD&C Act [21 U.S.C. 301], PHS Act, 21 CFR, and other acts where FDA has enforcement authority unless these cites require concurrence or are specifically prohibited – see IOM 5.2.3.3 Non-Reportable Observations. Examples of these observations generally fall into two categories.

5.2.3.2.1 - Adulteration Observations

Review Sections 402, 501, 505(K), 601, and 704 of the FD&C Act [21 U.S.C. 342, 351, 355(K), 361, and 374]. Include specific factual observations of:

1. Foods, drugs, devices, or cosmetics consisting in whole or in part of filthy, putrid, or decomposed substances.

2. Undesirable conditions or practices, bearing on filth or decomposition, which may reasonably result in the food, drug, device, or cosmetic becoming contaminated with filth.

3. Insanitary conditions or practices which may reasonably render the food, drug, device, or cosmetic injurious to health.

4. Careless handling of rodenticides or pesticides.

5. Results of field tests (organoleptic examination of fish, crackout of nuts, etc.) if the results revealed adulteration.

6. Observations of faulty manufacturing, processing, packaging, or holding, of food, drug, or device products as related to current good manufacturing practice regulations including inadequate or faulty record keeping.

7. Observations of faulty can closures and/or deviations from recommended processing times and temperatures.

8. Deviations from the animal proteins prohibited in ruminant feeds requirements (21 CFR 589.2000).

5.2.3.2.2 - Other Observations

You may include other factual observations of significant deviations from the FD&C Act [21 U.S.C. 301], 21 CFR, Government Wide Quality Assurance Program (GWQAP) requirements, and other Acts as directed by CPGMs and other agency directives. In some cases, you may cite labeling deviations as directed below. This list is not all inclusive.

1. Observations indicating non-conformity with commitments made in a New Drug Application (or NADA) or in am antibiotic certification or certification exemption form. See Section 505 FD&C Act, [21 U.S.C. 355].

2. Observations, forming the basis for product non-acceptance under the Government Wide Quality Assurance Program (GWQAP). See IOM 5.2.3.5.

3. Deviations from blood and blood products labeling requirements as specified in 21 CFR 606.121 and 21 CFR 640.

4. Animal protein products, and feeds containing such products, that are not in compliance with the labeling requirements of paragraphs (c) through (f) of 21 CFR 589.2000. See Section 403(a)(1) or 403(f) of the FD&C Act [21 U.S.C. 343(a)(1) or 343(f)].

5. Deviations from the applicable labeling regulations for human cells, tissue, and cellular and tissue-based products (HCT/Ps) as specified in 21 CFR 1271 and CPGM 7341.002.

6. Observations indicating drug misuse, failure to maintain proper drug use records, and/or poor animal husbandry practices during tissue residue investigations. See the applicable Compliance Program(s) for guidance.

7. Observations indicating non-conformity with the postmarketing adverse drug experience reporting requirements as specified in 21 CFR 310.305, 314.80, 314.98, 314.540, or 600.80 or other postmarketing requirements as specified in 21 CFR 314.81 or 600.14. See Sections 505 and 760 of the FD&C Act [21 U.S.C. 355(k) and 379aa].

8. Observations indicating non-conformity with the Medical Device Reporting requirements as specified in 21 CFR 803

   {See Section 519(a) of the FD&C Act [21 U.S.C. 360i]}; the Medical Devices Reports of Corrections and Removals requirements as specified in 21 CFR 806 {See Section 519(f) of the FD&C Act [21 U.S.C. 360i(f)]}; and the Medical Device Tracking requirements as specified in 21 CFR 821 {See Section 519(e) of the FD&C Act [21 U.S.C. 360i(e)]}.

9. Observations indicating noncompliance with medical device pre-market notification requirements and pre-market approval requirement under FD&C Act sections 510(k) and 515 [21 U.S.C. 360 (k) and 360e] respectively, should only be made with the prior confirmation of CDRH and/or CBER.

10. 21 CFR PART 200.10 does allow reporting observations noted at a contract facility to the contracting facility. Before doing this, check with your supervisor to determine if this is appropriate.

11. Observations indicating non-compliance with LACF/Acidified food registration and failure to file scheduled processes. Before doing this, verify lack of such, as covered in CPGM 7303.803A.

5.2.3.3 - Non-Reportable Observations

Do not report opinions, conclusions, or characterize conditions as "violative." The determination of whether any condition is violative is an agency decision made after considering all circumstances, facts and evidence. See IOM 5.2.7 involving discussions with management at which time opinions may be discussed.

Do not quote Regulations (e.g., specific CFR sections) when listing items.

Do not report observations pertaining to:

1. Label and labeling content, except per IOM 5.2.3.2.2, items 2, 3, 4 and 5 above.

2. Promotional materials.

3. The classification of a cosmetic or device as a drug.

4. The classification of a drug as a new drug.

5. Non-conformance with the New Drug Regulations, 21 CFR 312.1 (New Drugs for Investigational Use in Human Beings: Exemptions from Section 505(a)) unless instructed by the particular program or assignment.

6. The lack of registration required by Section 415 and 510 of the FD&C Act. The lack of registration per 21 CFR 1271 Subpart B Procedures for Registration and Listing, promulgated under Section 361 of the PHS Act.

7. Patient names, donor names, etc. If such identification is necessary, use initials, code numbers, record numbers, etc.

8. Corrective actions. Specific actions taken by the firm in response to observations noted on the FDA 483 or during the inspection are not listed on the FDA 483, but are reported in the EIR. Except as described in IOM 5.2.3.4.

9. The use of an unsafe food additive or color additive in a food product.

Use Turbo EIR to document in the "General Discussion with Management" section Non-Reportable Observations, which you discussed with management. These objectionable conditions fall into three basic categories:

1. Observations of significant deviations from specific Laws and/or regulations, non-reportable items 1-9 above.

2. Observations of deviations from specific Laws and/or regulations, which in your judgment, are of "questionable significance" and "deemed not to merit inclusion on the FDA 483," but do warrant discussion with management.

3. Observations which in your judgment deviate from official published guidance, not regulations, but warrant discussion with management.

The reporting of observations in these 3 categories is as follows:

Category 1: You should select the appropriate Turbo cite, verify or set the "Print type" to "Do Not Print," and save the observation in the Turbo database. This should be done even if there are no other reportable observations. For example, Lack of Food Registration as covered in IOM 5.4.1.5.3 is not reportable.

Category 2 or 3: You should always report these two categories of observations which were discussed with management under the "General Discussion with Management" heading in the EIR as specified by IOM 5.10.4.3.15. You have options in choosing how observations in category 2 are reported. You may select the appropriate cite in Turbo, enter the "specifically" text regarding the observation, and discussion with management, set it to "Do not print", save, and it will be automatically entered into the Turbo EIR when it is generated.

The second option which is also true for category 3 (i.e., there are no Turbo cites for official guidance, only regulations) is the observation/s discussed with management may be entered directly into the Turbo EIR under the "General Discussion with Management."

5.2.3.4 - Annotation of the FDA 483

Offer to annotate the FDA 483 for all medical device inspections. The district has discretion to annotate the FDA 483s in other program areas. BIMO inspections are generally excluded from annotations. Annotations of FDA 483s for inspections in other program areas may be done if both the establishment and the investigator/team believe annotation will facilitate the inspection process. When a FDA 483 is annotated it should be done in accordance with the guidance that follows.

Inform the establishment of the annotation program at some point prior to the final discussion with management. Determine from management whether they wish to have their FDA 483 observations annotated. It is voluntary on the part of the establishment. If the establishment does not want one or more observations annotated, you must honor the request.

The actual annotation of the FDA 483 should occur during the final discussion with management. The annotations are succinct comments about the status of the FDA 483 item. It is not permissible to pre-print or pre-format the annotations onto the FDA 483 form. The annotations can be made after each observation, at the end of each page of the FDA 483 or at the bottom of the last page of the FDA 483 prior to the investigator's signature. The establishment should review the annotations on the issued FDA 483 to ensure there are no misunderstandings about promised corrective actions. See IOM 5.2.3 for discussions of FDA 483 observations with management.

If the establishment has promised and/or completed a corrective action to an FDA 483 observation prior to the completion of the inspection, the FDA 483 should be annotated with one or more of the following comments, as appropriate:

1. Reported corrected, not verified.

2. Corrected and verified.

3. Promised to correct.

4. Under consideration.

   The term "verified" means "to confirm; to establish the truth or accuracy". In this case, you must do the verification. In some situations, you will not be able to verify the corrective action unless there is further district or Center review or until there is another inspection of the establishment.

The establishment's stated objections to any given observation or to the FDA 483, as a whole should not be annotated on the FDA 483. If they would prefer no annotation, do not annotate it. The EIR should include the establishment's objections to the observation and the fact the establishment declined to have the observation annotated.

When an establishment has promised corrections and furnishes a date or timeframe (without a specific date) for completion, then you may add "by xxx date" or "within xxxx days or months" in the annotation. Where the investigator and the establishment have "agreed to disagree" about the validity of an observation, you may annotate this observation with "Under consideration" or with no annotation based on the establishment's desire.

All corrective actions taken by the establishment and verified by FDA should be discussed in detail in the Establishment Inspection Report (EIR) and reported using the Compliance Achievement Reporting Systems (CARS).

5.2.3.5 - Government Wide Quality Assurance Program (GWQAP)

When performing product acceptance examinations under the GWQAP, you must discuss all deficiencies with management and report these deficiencies in writing on the FDA 483. This includes all deficiencies related to the FD&C Act as well as deficiencies in complying with contract requirements, which result in non-acceptance. There must be a clear differentiation on the FDA 483 between these two types of deficiencies.

Enter the FD&C type deficiencies (GMP deviations, etc.) first on the FDA 483. If there are deficiencies in contract provisions, draw a line across the sheet and add a heading "The Following Additional Contract Non-Conformances Were Observed." Enter each deficiency, which forms a basis for non-acceptance, followed by the reference to the applicable contract requirement or specification.

5.2.3.6 - Distribution of the FDA 483

Be sure all copies of the original FDA 483 are legible and distribute as follows.

5.2.3.6.1 - Original

The FDA 483 issued to the firm signed in pen and ink.

Before leaving the premises at the end of the EI present the original to the individual who received the FDA 482, Notice of Inspection, if the person is present and qualifies as "most responsible." If the person is not available or is outranked by someone else, present it to the individual who meets the definition of owner, operator, or agent in charge.

5.2.3.6.2 - Copies

Replicas of the "original".

Attach one copy of all FDA 483s issued to the firm to the EIR. This includes turbo or non-turbo copies of any signed, modified, and/or amended FDA 483, or 483 addenda. See IOM 5.2.3.1.6 (Correction of FDA 483 Errors). A copy may be sent to the top management of the firm including foreign management, unless the individual to whom you gave the original is the top official of the firm.

If the inspection covered vehicles as described in IOM 5.2.2.2, leave an exact copy of the list of observations with the firm being inspected. The original will be sent by your district to the firm owning or leasing the vehicle. You must make every effort to obtain the name and address of the vehicle owner. Usually the firm name is on the vehicle; however, it may require a trace of the vehicle license number. Discuss with your supervisor before taking this step. See IOM 4.4.7.2.

5.2.4 - Receipt - Factory Samples

You must issue an FDA 484, Receipt for Samples, if you collect any physical sample during an inspection. At the end of the EI and prior to leaving the premises, issue the original FDA 484 to the same individual who received the FDA 482. (See IOM 4.2.5) If this person is not available, give it to someone else who meets the definition of owner, operator, or agent in charge. Submit an exact copy with the EIR. Do not comment on type of examination expected or promise a report of analysis.

5.2.4.1 - Items Requiring Receipt

Issue FDA 484 for any item of food, drug, device, or cosmetic actually removed from the establishment.

NOTE: *A receipt must always be issued to anyone from whom you obtain Rx drugs. This includes individuals as well as firms. See IOM 4.2.5.4 and IOM 4.4.10.3.44.*

The following are examples of exhibit materials also requiring a Receipt for Samples:

1. Air filter pads,

2. Rodent pellets, and

3. Any other physical evidence actually removed from the plant.

5.2.4.2 - Items Not Requiring Receipt

Do not issue a FDA 484 for:

1. Items or materials examined during the inspection but not removed from the establishment (report adverse results of analysis of materials on FDA 483 as indicated in IOM 5.2.3.2),

2. Labels or promotional material, or

3. Photographs taken during the inspection.

4. Record(s): including production, quality control, shipping and interstate records.

Firm's management may request copies of documents or records you obtain from their firm. There is no objection to supplying them.

See IOM 5.3.8.5 for procedures when a firm requests a receipt for records copied during an inspection or investigation.

5.2.5 - Inspection Refusal

Refusal as used in your IOM means refusing to permit an inspection or prohibiting you from obtaining information to which FDA is entitled under the law. See IOM 4.2.3 for information regarding refusal to permit sampling.

In the case of a refusal you must show your conduct was reasonable, fair, and you exercised reasonable precaution to avoid refusal. You must have shown your credentials and given the responsible individual a properly prepared and signed Notice of Inspection, FDA 482.

Inspection refusals may take several forms. All refusals to permit inspection must be reported in your EIR under the "Refusals" heading.

5.2.5.1 - Refusal of Entry

When you are faced with a refusal of entry, call the person's attention to the pertinent sections of the Acts (Sections 301(f) and 704 of the FD&C Act [21 U.S.C. 331 (f) and 374] and Section 351(c), 360A(a), (b) and (f); 360B(a); and 361(a) of the Public Health Service Act. Portions of these are listed on the front and back of the FDA 482. If entry is still refused, leave the completed FDA 482, leave the premises and telephone your supervisor immediately for instructions.

5.2.5.2 - Refusal to Permit Access to or Copying of Records

If management objects to the manner of the inspection or coverage of specific areas or processes, do not argue the matter but proceed with the inspection. However, if management refuses to permit access to or copying of any record which you are entitled under law, call attention to Section 301(e) of the FD&C Act [21 U.S.C. 331] or applicable sections of the PHS Act. If management still refuses, proceed with the inspection until finished. It is not an inspection "refusal" when management refuses to provide formula information, lists of shipments, codes, etc., except where specifically required by the law. If the refusal is such you cannot conduct a satisfactory inspection, discuss with your supervisor if a Warrant for Inspection should be requested.

5.2.5.3 - Refusal after Serving Warrant

If you have been refused entry, obtained a warrant, tried to serve or execute it and are refused entry under the warrant, inform the person, the warrant is a court order and such refusal may constitute contempt of court. If the warrant is not then immediately honored (entry and inspection permitted), leave the premises and promptly telephone the facts to your supervisor.

If you have served the warrant and during the inspection you encounter partial refusal or resistance in obtaining access to anything FDA is authorized to inspect by the warrant, inform the firm that aspect of the inspection is part of a court order and refusal may constitute contempt of court. If the warrant is not then immediately honored, leave the premises and promptly telephone the facts to your supervisor.

5.2.5.4 - Hostile and Uncooperative Interviewees

More often than not, investigations or inspections are conducted in a reasonable atmosphere. Nonetheless, there will be times you are confronted by unfriendly or hostile persons.

Your activities must always be conducted with tact, honesty, diplomacy, and persuasiveness. Even though you must at times adopt a firm posture, do not resort to threats, intimidation, or strong-arm tactics.

Many times a hostile or uncooperative attitude on the part of individuals being interviewed results from fear, timidity, or previously distasteful encounters with law enforcement personnel. In most cases a calm, patient, understanding and persuasive attitude on your part will overcome the person's reluctance or hostility. Often the mere fact you patiently listen while individuals share their views will make them receptive to your quest.

5.2.5.4.1 - Indicators

Normally you have no way to predict the nature of the individuals you meet. However, there are often indicators, which can alert you, such as:

1. Establishment inspection reports, endorsements or memorandums may show situations where investigators encountered belligerent or hostile individuals. These reports may be FDA reports and/or State contract reports, if available.

2. Discussions and conversations with FDA, federal, state and local inspectors and investigators may reveal instances where uncooperative individuals and problem situations were encountered.

3. The nature of the assignment, program or information requested may indicate some degree of caution is needed.

4. A firm located in an area with a reputation for unfriendliness to law enforcement personnel should alert you some employees of the firm may be less than cooperative during the investigation.

If you find yourself in a situation which, in your judgment, indicates violence is imminent, stop the operation and make an exit as soon as possible. Immediately report the facts to your supervisor.

5.2.5.4.2 - Safety Precautions

The FDA recognizes there are situations where it is advisable to take precautions for your personal safety. In those, consult your supervisor. Some procedures, which may be utilized to minimize the danger, include:

1. Inspections or investigations carried out by a team of two or more persons.

2. Consider whether or not the use of an unmarked government car would be more beneficial to assist you in your inspection in lieu of a marked government car.

3. Request additional information from your State and/or Local Agencies who also regulate and inspect the facilities in question. In many instances, your State counterparts may have more information regarding the facility. This may be especially helpful for those firms that FDA has not yet inspected but were inspected by your State counterparts.

4. Each government car or inspection team should be assigned one FDA cell phone or alternate communication device. While we recognize that some Investigators carry a personal cell phone, FDA strongly suggests that your personal cell phone not be utilized to contact the firm or firm's management. In some instances, such uses in the past have resulted in later inappropriate contacts from the firm to the individual FDA Investigator.

5. Request assistance from local law enforcement agencies prior to or during investigations. This assistance may include information about the facility you are to inspect, assistance with communication devices, or police protection, if the police jurisdiction allows for such an action.

6. In potentially hazardous investigations such as methadone or schedule II Class Drugs, two investigators may be used and personnel from the U.S. Drug Enforcement Administration, State, or local law enforcement agencies may be requested to accompany you.

5.2.5.4.3 - Procedures when Threatened or Assaulted

In instances when you are actually assaulted or threatened, you should immediately notify your supervisor. Your supervisor can summon local police, United States

Marshals, or contact OCI headquarters for assistance (301-294-4030). OCI can make contacts with local police and federal agencies based on previous liaison. Also, the District should notify Division of Field Investigations, HFC-130 at 301-827-5653 FAX 301-443-3757 or via e-mail ORAHQDFICSOSAF@FDA.HHS.GOV.

If you are physically attacked, you have the same recourse as any other citizen as well as the benefit of federal laws protecting government officials while in the performance of their official duties. If you are physically attacked, you should get to safety, call your supervisor, report the incident and seek medical attention if needed. Remember that the medical attention you receive may be used as documentation for the Agency in support of any legal action taken against the firm or the individual.

5.2.5.4.4 - Notification of FBI and US Attorney

It is a federal crime for anyone to kill, assault, resist, oppose, impede, intimidate, or interfere with, a federal official in the performance of their official duties.

In case of assault or threat against you, notify your supervisor immediately, so the facts can be submitted to the Federal Bureau of Investigations and the U.S. Attorney's office for immediate action.

The referenced sections in Title 18 of the U.S. Code are:

1. Title 18 U.S.C.A. Section 111, which provides:

    "111. Assaulting, resisting, or impeding certain officers or employees.

    Whoever forcibly assaults, resists, opposes, impedes, intimidates, or interferes with any person designated in Section 1114 of this title while engaged in or on account of the performance of his official duties, shall be fined not more than $5,000 or imprisoned not more than three years, or both.

    Whoever, in the commission of any such acts uses a deadly or dangerous weapon, shall be fined not more than $10,000 or imprisoned not more than ten years, or both. **** ".

2. Title 18 U.S.C.A. Section 1114, which provides:

"1114. Protection of officers and employees of the United States.

Whoever kills ***** or any officer or employee of the Department of Health and Human Services or of the Department of Labor assigned to perform investigative, inspection, or law enforcement functions while engaged in the performance of his official duties, shall be punished as provided under sections 1111 and 1112 of this title. ****".

See Title 18 of the US Code Sections 111 and 1114 for the complete text. See also IOM 1.5.

5.2.6 - Inspection Warrant

A refusal to permit inspection invokes a criminal provision of section 301(f) of the FD&C Act [21 U.S.C. 331(f)]. Depending on the individual situation, instances of refusal may be met by judicious use of inspection warrants.

Instructions for obtaining warrants are contained in the Regulatory Procedures Manual, Chapter 6-3. See your supervisor for information and instructions.

You are operating as an agent of the court when you serve an inspection warrant and it must be executed expeditiously once served. See IOM 5.2.5.3 for guidance on how to handle any refusal after obtaining a warrant.

In situations where a potential problem is anticipated with the service of a warrant, the District should consider sending a Supervisory Consumer Safety Officer or Compliance Officer and a U.S. Marshal with the Investigator to assist and supervise the serving of the warrant.

After obtaining an Inspection Warrant, return to the firm and:

1. Show your credentials to the owner, operator, or agent in charge,

2. Issue the person a written Notice of Inspection (FDA 482),

3. Show that individual the original signed Inspection Warrant,

4. Give him/her a copy (not the original) of the warrant.

The copy you provide need not be signed by the issuing judge, but the judge's name should be typed on the copy.

Follow the procedures of the court or U.S. Attorney involved, if their methods differ from the above.

When an inspection is made pursuant to a warrant, a Return showing the inspection was completed must be made to the Judge (or U.S. Commissioner or Magistrate) who issued the warrant. The Return, executed on the original warrant, should be made promptly and usually no later than 10 days following its execution.

5.2.7 - Discussions with Management

After completion of the inspection, meet with the highest ranking management official possible to discuss your findings and observations. The FDA 483 is not a substitute for such discussion since there may be additional questionable practices or areas not appropriate for listing on this form.

During the discussion be frank, courteous and responsive with management. Point out the observations listed on the FDA 483, are your observations of objectionable conditions found during the inspection, and explain the significance of each. Try to relate each listed condition to the applicable sections of the laws and regulations administered by the FDA. You should inform management during the closeout discussion the conditions listed may, after further review by the Agency, be considered to be violations of the Food, Drug and Cosmetic Act or other statutes. Legal sanctions available to FDA may include seizure, injunction, civil money penalties and prosecution, if establishments do not voluntarily correct serious conditions. Do not be overbearing or arbitrary in your attitude or actions. Do not argue if management voices a different view of the FDA 483 observations, or of your opinions. Explain, in your judgment the conditions you observed MAY be determined by the FDA, after review of all the facts, to be violations. Make clear the prime purpose of the discussion is to call attention to objectionable practices or conditions, which should be corrected.

Obtain management's intentions regarding correcting objectionable conditions. They may propose corrections or procedural changes and ask you if this is satisfactory. If this involves areas where your knowledge, skill, and experience are such that you know it will be satisfactory, you can so advise management. Do not assume the role of an authoritative consultant. In areas where there is any doubt, you must explain to management you cannot endorse the proposed corrections. Advise the individuals FDA will supply comments (see RPM 4-1-3 #4) if the establishment will submit its request and its proposed corrections or procedures in writing to the district office.

Concentrate on what needs to be done rather than how to do it. Do not recommend the product or services of a particular establishment. If asked to suggest a product or consulting laboratory, refer the inquirer to a classified directory or trade publications and or organizations.

Report in your EIR all significant conversations with management or management representatives. In most instances it is not necessary to quote management's response verbatim. Paraphrasing the replies is sufficient. However, if the situation is such that quoting the reply or replies is necessary, enclose them in quotation marks.

5.2.7.1 - Protection of Privileged Information

You have certain responsibilities under the FD&C Act, Section 301(j); Sections 359(d) and 306(e) of the Public Health Service Act; and Section 1905 of the Federal Confidential Statute (18 U.S.C. 1905) regarding protection of confidential material obtained during your official duties. See IOM 1.4.

Do not volunteer information about other firms or their practices. Ignore casual exploratory questions or remarks from management about competitors or their processes. Your casual and seemingly innocuous remarks may reveal privileged information. Therefore, be alert and avoid voluntarily or unknowingly divulging information, which may be privileged or confidential and possibly compromise FDA's and your own integrity.

Management often request copies of any documents or records you obtain from their firm. There is no objection to your supplying these. When management requests copies of photos taken by you in a plant, follow IOM 5.3.4.5.

You may encounter situations when management invites outside individuals to observe the inspectional process (e.g., representatives from the press, trade associations, congressional staff, other company officials). As discussed in Section 5.1.4.3 of the IOM, the presence of representatives invited by the firm should not disrupt the inspectional process. You are to continue the inspection in a reasonable manner.

If the firm allows invited individuals to photograph, videotape, or prepare audio recordings during the inspection, you should make every effort to protect privileged information in your possession. However, it is the Agency's position that it is the firm's responsibility to protect confidential and/or proprietary information observed or recorded by those individuals invited by the firm. Where applicable, refer to IOM 5.3.5 for additional procedures on how to prepare your own recording in parallel with the firm's recording.

5.2.7.2 - Refusals of Requested Information

Should management refuse to provide any reasonable request for information, which is not specifically required by the law, determine the reasons for the denial and report the details in the EIR. Types of refusals of interest to FDA and refusal codes to be entered in FACTS are listed in the FDA Data Codes Manual. Refusal codes' data are used when reporting to Congress. See IOM 5.2.5.4 for instructions in dealing with hostile and/or uncooperative interviewees.

5.2.8 - Consumer Complaints

Prior to conducting any inspection, you should review the FACTS system and the factory jacket becoming familiar with all FDA Complaint/Injury forms. Be especially alert for ones marked "Follow-Up Next Inspection" and make sure you investigate these during your inspection.

During the inspection, discuss these complaints with management without revealing the complainant's name(s). Determine if the firm has had similar complaints on the same product. Determine what action the firm has taken to identify the root cause of the problem and to prevent a recurrence in the future. See IOM 5.10.4.3.11 for reporting instructions.

5.2.9 - Interviewing Confidential Informants

When you are faced with a situation involving sources of information who want to remain anonymous, please contact your supervisor and follow the procedures here. In addition, refer to IOM 5.2.1.2 regarding your personal safety. If your

management concurs with the decision to utilize a
confidential source, it is particularly important you take the
necessary steps to keep the identity of the source, and any
information which could lead to the identity, confidential. For
purposes of this subchapter, a confidential source is a
person who provides information that may be of assistance
to FDA without necessarily becoming a party to the actual
FDA investigation. If you believe the information provided
by the source could lead to a criminal investigation, please
contact the Office of Criminal Investigations (OCI).

5.2.9.1 - How to handle the first contact

When you interview a person who may become a
confidential source use the following procedures:

1. *Type of meeting.* Try to schedule a personal interview
   with the person rather than a telephone interview. At a
   face-to-face interview you can assess the person's
   demeanor, body language, overall presentation, and
   truthfulness.

2. *Meeting location.* The place and time of the interview
   should be the choice of the person, unless there is a
   concern with personal safety. If the person's suggested
   location is unsuitable, the investigator should suggest
   the location. When you conduct the interview off FDA
   premises, notify your supervisor of your destination,
   purpose, and estimated time of return. When an off-site
   interview has been completed, check-in with your
   supervisor.

5.2.9.1.1 - Interviewing methods/techniques

It is strongly recommended you have two investigators
conduct interviews of a confidential source. The lead
investigator conducts the interview, while the second
investigator takes notes and acts as a witness to the
interview. You should:

1. Prepare carefully for the interview. The investigators
   should develop the questions they intend to ask the
   person during the interview, e.g., "establish motivation,"
   and record and number the questions to be asked in
   their diaries prior to the interview. This preparation
   assists in documenting the interview process and
   reduces the amount of note taking needed during the

interview. The investigators also should discuss their interviewing strategy, and determine the method by which they will consult with each other during the interview and (during extensive interviews) share the interviewing and note-taking responsibilities;

2. Have the person tell the story chronologically, placing complex situations into logical order; and

3. If the person makes allegations, ask him or her how he or she knows the allegations are true.

    a. How were they in a position to know?

    b. Did they personally see, hear, or write about the information/incident?

    c. Can they provide proof of the allegations?

5.2.9.1.2 - Establish motivation

At the end of the interview ask the person why he or she is divulging this information. This may reveal their motive(s):

1. Is the person a disgruntled current or former employee who harbors a grudge?

2. Is the person looking for some type of whistle-blower reward or notoriety?

3. Does the person just want to do the right thing?

4. Is the person involved in actual or prospective litigation about or related to the information?

5.2.9.1.3 - Anonymity

If the person is requesting anonymity, inform him or her FDA:

1. Will not divulge his or her identity, the occurrence of the interview, or the sensitive information provided to FDA if the information could lead to the identity of the person, unless FDA is required to disclose the information by law, e.g., the investigation leads to a hearing or trial and he or she is required to testify, and

2. Will try to corroborate all information provided by the person, minimizing the chances he or she must later testify. However, testifying remains a possibility.

Ask the person for names of other persons who might be willing to speak with you about the allegations and corroborate their story.

5.2.9.2 - Protect the identity of the source

Collection of information. Obtain sufficient personal information necessary to enable you to contact the person for follow up if needed. However, to maintain the confidentiality of the person, do not include the person's identifier information such as gender, name, address, and phone number in the memorandum of interview. You should assign the confidential source a code name or number and use the identifier in memoranda and other communications relating to the confidential source (see IOM 5.2.9.2.2 item 2).

5.2.9.2.1 - Access

Know who is authorized by District procedure to access the information, and restrict access by others accordingly. Share the minimum amount of information necessary to meet the purpose of the disclosure.

5.2.9.2.2 - Storage Requirements

Each District should establish procedures, in addition to those listed below, to properly store confidential information. The following list contains information related to storage procedures.

1. Use security measures necessary to protect the confidentiality of personal information, whether it is in hard copy or electronic form, on FDA premises, in an FDA home-based computer, or in any other form. Use whatever means necessary and appropriate to physically safeguard the information, such as storing in a safe, or locked file cabinets, or password-coded computers, etc.

2. When referring to the source in any manner (orally, in writing, electronically, etc.), consider using code to identify the source. For example, use a number rather than the individual's name, to identify the source.

Personal privacy information should be safeguarded. Use discreet subject headers in the file labels as appropriate.

3. Remove personal information from a file only after you have noted in the file your name, date, etc. Promptly return that information to the file.

5.2.9.2.3 - Disclosure

Do not disclose information from or about the source, unless the disclosure complies with the law and FDA's procedures. Do not share non-public information outside of the Freedom of Information (FOI) process, unless the sharing is done according to our regulations and procedures. Refer FOI requests to your FOI officer (see item 3 below). See also IOM Subchapter 1.4. The following information relates to disclosures of information from or about a confidential source.

1. Make duplicates of the personal information only to the extent necessary for authorized disclosure (inside or outside of FDA). Do not leave the copy machine unattended.

2. Make only authorized disclosures of the information, regardless of the manner of disclosing (oral, written, etc.). Do not use mobile telephones or leave voice mails with the information. Avoid transmitting the non-public information by facsimile or e-mail.

3. If you receive a FOI request for information from or about a source consult with your supervisor immediately Disclosure to a non-FDA government official of information from or about a source may be disclosed only if permitted by law and FDA procedures, and after consulting your supervisor and, if needed, OCI.

4. Immediately retrieve information from or about a source is inadvertently disclosed.

5.2.9.2.4 - Destruction

Destroy personal information by shredding or similar means which physically destroys the record and/or, if the information is in electronic form, makes it unreadable.

Office of Chief Counsel. After a matter has been referred to the Office of Chief Counsel (OCC) for litigation or enforcement action, consult with OCC if you are interested in contacting the source.

5.2.10 - Routine Biosecurity Procedures for Visits to Facilities Housing or Transporting Domestic or Wild Animals

This section is FDA's guidance when you visit any type of facility where any domestic or wild animals are housed or transported. If a firm has more restrictive controls, follow those in addition to the controls cited below as long as they do not interfere with your assignment needs. The controls and procedures are intended to prevent you from becoming a vector or carrier of animal diseases, to prevent the spread of animal disease, and to set a good example for stockmen, growers and industry servicemen. A number of chronic diseases, such as Johne's Disease, bovine virus diarrhea (BVD) and others exist in domestic animals which you can unknowingly spread. Any inspectional contact with herds of livestock (including poultry) or non-domesticated animals exposes you to potential claims of introducing or spreading disease. This could occur between sections of a single site, such as poultry houses, or between different sites or farms. The potential also exists for the introduction of disease from an animal processing plant, such as a slaughterhouse or renderer to a live animal facility. You can prevent this by following appropriate cleaning and disinfection steps between facilities. Generally, a break of 5 days or more between sites is sufficient to eliminate concern about transmission of infectious agents.

These precautions, biosecurity measures, are necessary in two types of situations. The first is when there is no known disease present and your actions are precautionary. This section primarily addresses those kinds of activities. The other situation involves known or suspected disease outbreaks or more notorious disease conditions such as salmonella in eggs, infectious Laryngotracheitis, foot and mouth disease, vesicular stomatitis, and blackhead which can be highly contagious and spread from one group of animals to another by movement of people and objects between infected and non-infected groups. In these cases, special precautions must be taken to make sure you are not an unknowing vector for the spread of disease. See IOM 5.2.10.3.

If you will only be inspecting an office or house away from areas where animals are housed or kept, clean and suitable street attire may be sufficient. Be aware if you visit any area of a facility where animals have been, you should always sanitize, clean or change footwear and it may be necessary to change outerwear before visiting another animal site to prevent any possibility of transmission of disease.

Your vehicle may also transport infection if you drive through contaminated areas.

5.2.10.1 - Pre-Inspection Activities

When you know you are going to visit or inspect any animal production or holding facility, consider contacting the State Veterinarian and/or the Regional APHIS office to determine if there are any areas in the state under quarantine or special measures to control animal diseases. APHIS office locations can be found on their website. The State Veterinarian will be listed under Government Listings in your phone book and is listed at this website. Regional Milk Specialists frequently working with State counterparts in the Interstate Milk Shippers program should contact these sources at least quarterly for updates. Ask for any special controls or procedures they recommend. Follow any guidance they offer in addition to the precautions in this section. You should also consider pre-notification of the facility following guidance in IOM 5.2.1.1, Pre-Announcement, unless your assignment does not allow pre-notification. If you elect to pre-announce the inspection, in addition to the normal contact, ask to speak with the person at the facility responsible for their biosecurity measures and find out what they require of employees and visitors. If their requests do not interfere with your ability to do your job, follow their requests as we do when inspecting sterile manufacturing facilities.

Make sure your vehicle is clean and has been recently washed. Commercial car washes are adequate as long as you check to make sure any dirt, manure or other debris, which may be present from a previous site, has been removed. Some facilities may require additional disinfection of tires upon entry to the premises. Ensure tires and floor mats are clean. Consider designating places in your vehicle for storage of clean, unused supplies and dirty or used supplies.

In addition to your normal inspectional tools, obtain the following equipment and supplies from your district:

1. Laundered or disposable coveralls or smocks (coveralls are suggested because they give better coverage). If you are going to visit multiple facilities in one day or trip, obtain sufficient quantities so you can change into clean or unused clothing between each site.

2. Disposable plastic gloves, rubber boots, which can be sanitized, and disposable shoe/boot covers. Rubber boots over which you place disposable shoe/boot covers are preferred.

3. Reusable cloth or plastic laundry bag(s) for clothing to be laundered. (Disposable bags can be used.)

4. Soap, water and disposable or freshly laundered individual hand (or paper) towels.

5. Sanitizing solution(s) and equipment (brushes, bucket, tray, measuring devices, etc.) to permit you to properly sanitizing hands, boots, equipment and your vehicle. Most disinfectants will require removing organic matter before use and good brushes are essential to remove dirt from boots and other objects.

Make sure any equipment you take with you has been thoroughly cleaned and sanitized as necessary. Clip boards, briefcases, flashlights, inspectional sampling tools, coolers, brushes, buckets and other objects should be cleaned between uses as necessary and between visits to any suspected infected facilities. Disposable equipment should be used to the fullest extent possible.

Maintain copies of any applicable Material Safety Data Sheets (MSDS) for disinfectants with you in your vehicle. If the firm's management requests information on the disinfectants you are using, they may read or copy these MSDS. Be familiar with the instructions and precautions concerning use of disinfectants. Any disinfectant should be effective against known or suspected microbiological agents.

In the event of a foreign animal disease, contact the USDA, APHIS Veterinary Services area Veterinarian in Charge for additional precautions and procedures to follow. (See 5.2.10.3)

5.2.10.2 - General Inspection Procedures

Always begin each day with a clean vehicle free from any visible dirt or debris. During the day, take precautions to minimize contamination of your vehicle. If your vehicle becomes obviously dirty with adhering mud or manure, clean it before visiting another animal facility. When you arrive at a facility where animals are located, check to see if there are designated parking spots or pads for visitors. If so, park your vehicle there unless directed otherwise by the firm. If there is no guidance, park well away from all areas housing animals. When you arrive, inquire about or reconfirm any biosecurity measures the firm employs. Confirm your actions are suitable and follow expectations of the facility when this does not interfere with your inspection ability. Follow steps requested by the firm to remove contamination from vehicles, which may include troughs or pools of disinfectants for tires or other control measures. Avoid driving through manure, mud or wastewater at these sites.

In general, entry to animal housing or feeding areas, corrals, calf pens, hospital pens or special treatment facilities should be avoided unless the assignment requires their inspection or there are specific reasons requiring entry. If you must visit the feeding area occupied by livestock or birds, first determine if any groups are infected with disease. Arrange to visit the known non-disease areas first. Do not handle any animals unless official duty requires such contact. Before leaving the area where you parked your car, put on protective clothing as described and proceed with the purpose of your visit; sanitizing hands (and gloves if worn) and boots as necessary during the visit or inspection.

General procedures:

1. Wear rubber boots or other suitable footwear, which you disinfect upon arriving at the site and prior to departure. It is preferable to also place disposable foot coverings over your footwear, regardless of the type, after you have disinfected them. If the firm has footbaths, use them. Boots and footwear should be disinfected with any of the agents identified at the end of this subsection using a good brush. Clean and disinfect the brush(es) and bucket you use for these activities.

2. Wash your hands with soap and water. If you are visiting a facility where a known animal disease is present or the firm's biosecurity protocol requires, wear disposable gloves.

3. Wear disposable or freshly laundered coveralls, when appropriate. Some facilities may provide disposable coveralls and require visitors to shower in and shower out at their facilities. If requested by the firm and facilities are provided, you should follow those requests.

4. Wear appropriate head coverings, as necessary. If you wear a head covering, clean and disinfect between facilities or use disposable head coverings.

5. Minimize any materials you carry with you such as notebooks, flashlights, etc. to what is required. Consider keeping these things in clean plastic bags or containers between uses. Disinfect any of these types of items as best you can between visits to facilities or between different animal-housing areas.

6. If you are visiting production units with animals of multiple ages, always try to work from the youngest to the oldest.

7. Avoid direct contact with livestock or wild animals, bodily fluids or animal byproducts when visiting facilities.

8. Regional Milk Specialists, Milk Safety Branch and State Training Team staff frequently working with State counterparts in the Interstate Milk Shippers program shall follow any biosecurity measures the firm employs, any biosecurity measures the State employs, and as a minimum shall follow the coded memoranda issued by CFSAN Milk Safety Branch on this subject.

Upon completing your assignment in a given animal area, return to the same area where you donned protective clothing. Remove disposable shoe/boot covers and gloves, if applicable, and place them in a disposable paper or plastic laundry bag. Clean and sanitize boots/footwear. Remove the protective clothing, if applicable, by peeling it off inside out. (This keeps the surfaces exposed to contamination on the inside.) Place all disposable items in a disposable laundry bag for disposal back at your office. If the firm has special containers for disposing of such articles, it is preferable to leave them there rather than transport them back to the office. Place reusable coveralls or other reusable protective clothing in a separate laundry bag for disposition at the office.

Follow guidance on biosecurity provided in the applicable Compliance Program or "Guide to the Inspection of '***'" in addition to precautions in this Section.

Repeat these procedures for each separate location visited or inspected.

Purchase commercially available solutions for disinfecting objects or consult with your servicing laboratory. Commercial products such as Nolvosan, Efersan, One Stroke Environ or Virkon-S may be used as long as they are registered by EPA for the intended purpose. Lye or chlorine based cleaners and disinfectants may also be used.

The following formula for household bleach may be used. Mix 3/4 cup (6 oz) of liquid bleach (5.25%) in one gallon of water (128 oz). This solution will be approximately 1:20 dilution. Formulations of household bleach, which are more concentrated than 5.25% are commercially available. Dilute accordingly to these directions. A more concentrated 1:10 solution (1-oz bleach to 9-oz water) may be used with decreased contact time required. Dilutions should be prepared fresh daily and protected from light.

You should read the label and be familiar with directions and precautions, such as removing any organic matter from objects to be disinfected, for any disinfectant you use. In the absence of directions or for chlorine solutions you prepare: 1. Remove visible dirt from the object (boots, tools, tires, etc.). 2. Wipe, brush or scrub surfaces with the solution and keep wet for 2 minutes. 3. Allow to air dry or dry with previously sterilized toweling.

5.2.10.3 - Special Situation Precautions

If you are required to inspect or visit a facility known or suspected to be involved in a contagious animal disease an outbreak or otherwise identified as having diseased animals, contact the Center for Veterinary Medicine and/or Center for Food Safety and Applied Nutrition for additional precautions which may be necessary before you visits these sites. Your activities may be limited to visiting a single site in a day, taking extra-ordinary decontamination steps, ensuring you do not visit or inspect another facility for 5 or more days following the visit to the contaminated site or other steps. APHIS may have special restrictions or precautions for you to follow. The State Veterinarian may also request you follow additional requirements. During inspections of poultry operations where salmonella contamination is known or suspected, you should make sure you contact CFSAN directly for specific procedures to follow. Additional decontamination steps will be required.

5.3 - Evidence Development

5.3.1 - Techniques

The recognition, collection, and effective presentation of admissible evidence is essential to successful litigation. Tangible evidence is required to support your observations and reports of violative conditions.

Although the inspectional procedures to detect adulteration and contamination, etc., are described under specific headings in the IOM, the same procedures and/or techniques may also apply to other areas. For instance, the procedures to detect contamination from filth, insects, rodents, birds, etc., described in IOM section 5.4.7 may also apply to drugs or other products. Your experience and training assists you in making this transition and enables you to detect possible violative conditions.

Keep in mind the policy annunciated in the 4/23/1991 memorandum from the Director, Office of Compliance: The lack of a violative physical sample is not a bar to pursuing regulatory and/or administrative action providing the CGMP deficiencies have been well documented. Likewise, physical samples found to be in compliance are not a bar to pursuing action under CGMP charges.

5.3.2 - Factory Samples

Samples of raw materials or finished products collected during inspections provide the necessary key to establish routes of contamination. They also document the character of products packed prior to the inspection. Collect Factory Samples for laboratory examination only when they contribute to confirming the suspected violation. Be selective since negative reports of analysis of food samples are required under Section 704(d) of the FD&C Act [21

U.S.C. 374 (d)] and might give management a false picture of the firm's operation.

When possible collect duplicate subsamples to provide for the 702(b) portion of the sample. See IOM 4.3.2.1 and 4.3.7.4.1 for additional guidance and 21 CFR 2.10 for exemptions regarding the collection of duplicate portions.

5.3.3 - Exhibits

Impressive exhibits are extremely effective and important forms of evidence to establish existence of violative conditions or products. They should relate to insanitary conditions contributing or likely to contribute, filth to the finished product, or to practices likely to render the product injurious or otherwise violative. Diagrams of the establishment, floor plans, flow charts, and schematics are useful in preparing a clear concise report and in later presentation of testimony. A small compass is useful in describing exact locations of objectionable conditions in the plant, in your diagrams, and locations from which samples were taken, etc. See IOM Exhibit 4-5.

Describe and submit under one INV Sample Number all exhibits (except photographs) collected during the inspection or investigation. Identify and number individual subs and officially seal all samples collected.

Examples of exhibits include:

1. Live and dead insects.

2. Insect frass, webbing, and insect chewed materials; nesting material of rodents and/or other animals; and other behavioral evidence of the presence of insects, rodents and other animals.

3. Samples of components or ingredients, in-process materials and finished products or dosage forms.

4. Manufacturing and control devices or aids.

5. Physical samples if possible and practical or, photographs with descriptions of scoops, stop-gap expediencies, other unorthodox manufacturing equipment or makeshift procedures. If photos are taken, follow the procedures described in IOM 5.3.4.

6. Evidence showing the presence of prohibited pesticide residues. A method of swabbing for prohibited pesticide residues was published in Laboratory Information Bulletin # 1622. Excerpts are quoted as follows:

    a. *Apparatus* - Four dram size glass vials, 95% ethanol, and cotton swabs preformed on 6" long wooden handles. Keep uncontaminated in a clean plastic bag.

    b. *Procedure* - Blow away loose dirt or debris from approximately a 3" x 3" selected area. Measure approximately 2 cm of 95% ethanol in vial, dip swab into ethanol, press out excess on inside of vial and roll moist swab back and forth firmly across the selected area. Return swab to vial, swirl in alcohol, press out excess on inside of vial and again roll moist swab across the same area 90° to the previous swabbing. Re-insert swab into vial, break off swab handle and cap the vial with the swab inside.

    c. When swab subsamples are submitted, also submit a blank control sub consisting of an unused swab placed in a capped vial containing 2 cm of the same alcohol that was used for the other swabs.

    d. Describe the type of material swabbed (cardboard carton, metal table top, rubber inspection belt, etc.) and the area covered. A reasonable area is approximately 10 sq. inches. Always try to establish a definite link in the chain of subsamples leading towards the highest level of contamination. If possible, identify the pesticide suspected. Be sure to include a floor plan with the areas sampled identified.

5.3.4 - Photographs

Photos taken during EI's are not classified as INV Samples. They are exhibits. No C/R is used for photos taken unless the photos are part of an Official Sample. See IOM 4.1.4 for information on Official Samples.

Since photographs are one of the most effective and useful forms of evidence, every one should be taken with a purpose. Photographs should be related to insanitary conditions contributing or likely to contribute filth to the finished product, or to practices likely to render it injurious or otherwise violative.

CAUTION: Evaluate the area where flash photography is contemplated. Do not use flash where there is a potentially explosive condition; e.g. very dusty areas or possible presence of explosive or flammable vapors. In these situations use extremely fast film and/or long exposure time instead of flash.

Examples of conditions or practices effectively documented by photographs include:

1. Evidence of rodents or insect infestation and faulty construction or maintenance, which contributes to these conditions.

2. Routes of, as well as, actual contamination of raw materials or finished products.

3. Condition of raw materials or finished products.

4. Employee practices contributing to contamination or to violative conditions.

5. Manufacturing processes.

6. Manufacturing and various control records showing errors, substitutions, penciled changes in procedure, faulty practices, deviations from GMP's, NDA's, or other protocols, altered or inadequate assays or other control procedures and any variation from stated procedure. See IOM 5.3.8.1 for identification of records.

7. Effluent contamination of water systems. See IOM 5.4.3 for techniques in photographing this type of contamination.

When photographing labels, make sure your picture will result in a legible label with printing large enough to be read by an unaided eye. Photograph whited out documents by holding a flashlight against the whited out side and taking a close up photo of the reverse using high-speed film. This will produce a photo with a mirror image of the whited out side.

If you use a Polaroid camera or color slide film, explain the facts in your EIR or on the C/R to alert reviewers that there are no negatives.

5.3.4.1 - In-Plant Photographs

Do not request permission from management to take photographs during an inspection. Take your camera into the firm and use it as necessary just as you use other inspectional equipment.

If management objects to taking photographs, explain that photos are an integral part of an inspection and present an accurate picture of plant conditions. Advise management the U. S. Courts have held that photographs may lawfully be taken as part of an inspection.

If management continues to refuse, provide them with the following references:

1. "Dow Chemical v. United States", 476 U.S. 227 (1986) This Supreme Court Decision dealt with aerial photographs by EPA, but the Court's language seems to address the right to take photographs by any regulatory agency. The decision reads in part, "** When Congress invests an agency with enforcement and investigatory authority, it is not necessary to identify explicitly each and every technique that may be used in the course of executing the statutory mission. ***"

2. "United States of America v. Acri Wholesale Grocery Company, A Corporation, and JOSEPH D. ACRI and ANTHONY ACRI, Individuals", U.S. District Court for Southern District of Iowa. 409 F. Supp. 529. Decided February 24, 1976.

If management refuses, advise your supervisor so legal remedies may be sought to allow you to take photographs, if appropriate. If you have already taken some photos do not surrender film to management. Advise the firm it can obtain copies of the photos under the Freedom of Information Act. See IOM 5.3.4.5.

5.3.4.2 - Photo Identification and Submission

One of the most critical aspects about photographs or videotapes is the ability for the agency to provide testimony clearly verifying the authenticity of the conditions depicted in the photograph or video. It makes no difference if the photo is a 35 mm print from acetate negatives, a Polaroid photo, a digital photo or video taken with a video recorder. You must create a trail, starting with the taking of the photo, confirming its original accuracy and establishing a record describing the chain of custody. To do this, you must make sure each photograph is described in your regulatory notes in sufficient detail to assure positive correlation of the photo or video with your inspection findings. One way you can do this is to photograph a card with your name, district address and phone number as the first frame or picture on a roll of film or in the digital record. This will help identify the film or file and assist in tracking if it is lost or becomes separated from its identification envelope during processing or storage. Proper procedures will also allow the agency to provide evidence confirming the authenticity of the photographs or video recording in the event you are not able to testify personally.

5.3.4.2.1 - Prints

Identify each print on the margin with exhibit number, firm name (or DOC Sample Nos., if DOC Sample), date taken or inclusive dates of inspection, and your initials. Do not place any identifying marks on the picture area of the print. (Some photo developing firms are supplying borderless prints. For this type print, place identification along the back bottom edge of the print and mount the print so the identification can be read without removing the print from the mounting paper. Place a narrative description on the mounting paper next to the print and attach as exhibits to the EIR and/or route with other records associated with a DOC Sample.)

5.3.4.2.2 - Color Slide Identification

If color slides are used, identify each slide, in the same manner as for prints. Districts may have special mounting frames for color slides, so the narrative description of each slide must be in the body of the report with proper reference to exhibits, or, each description may be placed on sheets of paper following the mounting frames and properly referenced.

5.3.4.2.3 - Negative Identification

Identify the edge of at least two negative strips, with the same information as for prints using a 3/16" strip of pressure sensitive tape. Place all negatives in a FDA-525 envelope. Complete blocks 2, 3, (4 if DOC Sample), 5, 7, and 12 and seal with an Official Seal, FDA-415a. If negatives are not part of a DOC Sample, enter firm name in the Sample Number block.

As applicable, submit the sealed FDA-525 or envelope as an exhibit to the EIR, with the Investigative Report as an attachment, or with the other associated records/documents with a DOC Sample.

5.3.4.2.4 - Video Recordings

Handle and protect the original video record just as if it were a photograph negative. Unused videotapes should generally be used to capture the video and, for subsequent copies of the original recording. Write-protect and identify the original videotape with a label with the firm name (or Sample number if it is being submitted as part of an official sample), date taken, and your initials. Officially seal the original videotape in a FDA-525 envelope or similar envelope. If you use a larger, unfranked envelope, identify the envelope with your name, title, home district, date, firm name, firm address (include zip code), description of the contents of the envelope, and marked in large, bolded letters "STORE AWAY AND PROTECT FROM MAGNETIC FIELDS." You may place more than one videotape in a single FDA-525 as long as you state on the envelope how many videotapes are in the envelope. If the original envelope is opened, document the chain of custody and use new seal(s) after each entry to the envelope.

If you perform any editing of the recording, you should only perform this on a copy of the original video recording to prevent possible damage to the original. Document in your regulatory notes you made a copy of the original and verified the copy is an accurate copy of the original video you took. This "original copy" should be treated just as if it is the original. When you sign the report, memorandum or other agency document, your signature certifies you are saying the content of the document, including any video recordings, is true and accurate to the best of your ability.

As applicable, submit the officially sealed FDA-525 or envelope as an exhibit to the EIR, with the Investigative Report as an attachment, or with the other associated records/ documents with a DOC Sample.

5.3.4.2.5 - Digital Photographs or Video Recordings

Prior to the year 2000, FDA investigators have traditionally worked with silver acetate photographic film or used analog video tapes. Early digital cameras recorded photographic images directly to floppy disks or mini CD-Rs in which the evidence could be handled like photographic negatives.

The important difference today is digital cameras are capable of recording high resolution images on the order of ten megapixels. The corresponding image file sizes can be over ten megabytes when using uncompressed file formats. To cope with the increased file sizes, digital camera manufacturers have introduced non-volatile flash memory cards which can record digital images, delete images, and be recorded over and over again. This presents a new issue since the original digital images, which are captured at the moment when the images are recorded on the memory card, will be copied at a later time to a CD-R or other permanent storage media. Due to the cost of flash memory cards and the large file sizes, it is not feasible to purchase new memory cards for each inspection/investigation as you did using photographic film. You will be working with a "original copy" of the images which have to be copied in the exact format to a CD-R as they were originally recorded on the flash memory card to preserve the chain of custody.

In the same manner, digital video recordings may involve the use of different media types such as tapes, CD-Rs or DVD-Rs, or built-in hard drives. If you cannot handle the original video recording as in IOM 5.3.4.2.4, you will need to create an "original copy" of the video recording.

Despite the differences in photographic film and digital technology, you are responsible for collection, handling, documenting the chain of custody, storage, and submission of your evidence in a manner where you can testify to its authenticity in a court of law. See IOM 5.3.4.2 and 5.3.4.3.

5.3.4.2.6 - GLOSSARY OF DIGITAL TERMINOLOGY

Some basic terminology is used when referring to digital devices in IOM 5.3.4.2.4, 5.3.4.2.5, 5.3.4.3.

5.3.4.2.6.1 – Digital Data

Electronic data in binary form consisting in its simplest form as "1"s and "0"s. A computer interprets data by whether the state is on ("1") or off ("0").

5.3.4.2.6.2 – Analog Data

Information captured in a directly measurable signal versus an analog signal converted and stored in binary.

5.3.4.2.6.3 – Memory Card

Any non-volatile memory media that can be removed and which retains data without the need for electrical power. Examples of current memory cards are: Compact Flash (CF), Secure Digital (SD), Memory Stick (Sony), and Extreme Digital (xD).

5.3.4.2.6.4 - Original

The file recorded by a digital device on digital storage media at the moment in time when the user takes a picture or makes a recording. This concept is similar to a film camera where the photographic film records the image when exposed by light. The film image negatives produced when the film is developed are considered the originals and prints are considered copies. See IOM 5.3.4.2.1 and 5.3.4.2.3.

5.3.4.2.6.5 – Original Copy

An exact copy of the original file recorded by the digital device (camera, video recorder, etc.). The original copy will retain all the characteristics of the original and is indistinguishable from the original.

5.3.4.2.6.6 - Permanent Storage Media

A media format in which the digital files cannot be altered once written. Examples are CD-Rs and DVD-Rs.

5.3.4.2.6.7 - Time/Date Stamp

The internal clock within the camera which records the time/date information on the image file. Set the time/date stamp for the location where the photographs or videos are being taken. In this usage, the time/date stamp does not refer to imprinting the time/date stamp within the photographic image although the time/date stamp can also be imprinted on the photograph as some film cameras could do.

5.3.4.2.6.8 - Working Copy

A copy of the original copy used when you need to make additional copies for your report, sample C/R. Creating a working copy decreases the chance the original copy is damaged.

5.3.4.3 - Preparing and Maintaining Digital Photographs as Regulatory Evidence

Assure and protect a digital photo's chain of custody (and authenticity) following this procedure:

1. Prior to using the digital camera, verify the date and time stamp is correct and there are no images stored on the memory card. Reformat the memory card using your camera's reformat command to delete any images not related to your current assignment. Depending on your inspection/investigation, camera, and memory card capacity you should consider bringing more than one memory card if possible.

2. Handle your camera and the memory cards in a manner to protect your evidence and maintain the trail of the "chain of custody" for the evidence you have collected. For example, keep the camera and memory cards in your personal possession at all times or held under lock/key in a secure storage area. Also, keep any additional memory cards containing images in your personal possession until transferred to permanent storage media. Where necessary, document these facts in your regulatory notes or written report (EIR, CR etc).

3. As soon as practical, create an original copy of the digital photos. Some older FDA cameras will capture images directly to a (Write-once Compact Disk Recordable (CD-R)); in this case, the CD-R from these cameras becomes the original CD-R. Identify, date and initial the CD-R as an original image record. If a CD-R/W was used, copy the images to a CD-R to create an original copy with files that can not be altered. Follow additional instructions for creating and finishing a CD-R in step 4 below.

4. If the camera requires downloading of images to a CD-R, download all the images from the digital camera to an unused CD-R or other electronic storage media to

create an original copy. If there was more than one memory card used, use a separate CD-R for each memory card. The storage capacity of a CD-R is about 650 mb; thus, more than one CD-R may be needed to create an original copy of your memory card depending on your camera's resolution, the storage capacity of your memory card, and the number of pictures taken. The images should be transferred in a file format maintaining the image resolution at the time the image was captured. If possible, avoid the use of any file compression in transferring the images to the CD-R. Prior to preparing the CD-R or transferring image files, verify that the computer you are using is set to the correct date and time. Make the CD-R permanent in a format readable by any CD-R reader. Prior to making the working copy from the original copy, identify the original copy with the same information as in IOM 5.3.4.2.1. It is important to identify the original copy as soon as possible to prevent possible mix up of the original copy with any working copies.

5. Use a permanent CD safe marker to identify the original copy CD-R. Do not use ball point pens or similar tipped markers since the CD-R may be damaged. See www.itl.nist.gov/iad/894.05/docs/CDandDVDCareandHandlingGuide.pdf. This NIST document, "Care and Handling of CDs and DVDs - A Guide for Librarians and Archivists" figure 12, page 23 shows where to identify the CD-R.

6. Where applicable, document in your regulatory notes the verification and identification of each photographic image comparing them to your regulatory notes, which were recorded at the time the photographs were taken.

7. Make only one working copy from each original copy and make any additional working copies using the initial working copy. No more than one copy should be made from the original copy in order to preserve the original copy a pristine set. After making the initial working copy, place the original copy in a suitable package, officially seal and store the original copy (CD-R or other electronic storage media) until submitted with the written report (EIR, CR etc). If the images are captured or transferred to diskettes, refer to IOM 5.3.8.3 for the handling of diskettes. If possible, the investigator (who took the photos and will authenticate them at trial)

should store the sealed CD-R or other electronic storage media until it is submitted with the written report. If you break the seal for any reason, document this on the broken seal, in your regulatory notes or written report, and reseal the package with a new official seal.

8. Working copies should be used to print photos, insertion into an EIR, cropped, otherwise edited or to be included in a referral.

9. Document in your regulatory notes or written report (EIR, CR, etc.) any steps taken for any unusual editing of original photo images. For example: Superimposing over a important area of the image, image enhancement, composite images, etc.

5.3.4.4 - Preparing Digital Photos for Insertion in a Turbo Establishment Inspection Report (EIR)

Digital photos taken during an inspection can be inserted into the body of a report in Turbo EIR or can be printed and attached to the EIR as an exhibit. Inserting digital photos can dramatically increase the file size of the Turbo EIR document. To maintain a minimum Turbo EIR document file size, the following is recommended: Do not open a digital picture/photo and use copy and paste to insert the picture/photo into the Turbo EIR document. Instead, save pictures/photos in a JPEG image format (.jpg file name extension) in a separate folder in preparation for inserting into Turbo EIR. Then resize all the JPEG pictures to a reasonable image file size. To do this,

1. Open the folder with all the pictures that may be inserted into the Turbo EIR document.

2. Hold the control key down and left click to select each image file to be resized.

3. Right click, choose resize pictures. See exhibit 5-6.

4. Select a size-- click on Small (fits a 640 x 480 screen), and click OK. Selecting one of the other screen sizes will also work with the exception of "Handheld PC (fits a 240 x 320 screen)"

5. New resized files will be created within the same folder. Each original file will be maintained. Each new resized file will be renamed as original file name (Small).jpg to differentiate it from the original file.

6. The resized pictures/photos are now ready for insertion into the Turbo EIR document. Remember to maintain the original image files and not the resized digital image files for submission with your hard copy report, forms, and exhibits to the official establishment file.

To insert a picture into the Turbo EIR document:

1. Open the Turbo EIR document Position cursor to where you want to insert the picture.

2. From the menu bar, click on Insert, choose Picture, click on From File, find and select folder with resized pictures to be inserted. See exhibit 5-7.

3. Double click on the resized picture to be inserted.

4. Picture inserted into the Turbo EIR document can be made larger or smaller by clicking on the picture and grabbing the corner of the picture frame and dragging to achieve the desired size.

 Captions can be added outside the borders of the picture or can be inserted within picture using more advanced photo editing techniques.

Alternative method: The Microsoft Office 2003 Tools folder contains a program called Office Picture Manager which can be used to resize pictures. See exhibit 5-8 which shows the "resize" menu option.

NOTE: *When any digital photos are used in an EIR, submit the original or original copy of the camera images following procedures as outlined in IOM 5.3.4.3 – Preparing and Maintaining Digital Photographs as Regulatory Evidence.*

5.3.4.5 - Photograph Requests

Do not routinely advise firms they may have copies of photos. However, if management of the firm initiates the request, advise them it is possible to obtain copies of photographs taken in their plant under the Freedom of Information Act. Any request should be sent to The Food

and Drug Administration, at the address listed on the FDA
482 or FDA 483. The firm must bear the cost of duplicating
the photographs.

Since photographs are records in an investigative file, they
are not available under the Freedom of Information Act until
the file is closed.

Do not discourage firms from taking their own photographs
at the same time and of the same scenes as you.

5.3.5 - Recordings

Under normal circumstances recording devices will not be
used while conducting inspections and investigations.
However, some firms are now recording and/or video
taping, the inspection and/or the discussion with
management portion of the inspection. These firms should
be advised we do not object to this procedure, but we will
also record the discussion to assure the accuracy of our
records. Occasionally a firm's management may record the
serving of an inspection warrant or, in a hostile situation,
may want to record everything. In such cases, depending
on the circumstances, you may prepare your own recording
in parallel with the firm's recording. Do not depend on the
firm to provide a duplicate of their recordings.

Use a clear tape cassette and identify the tape verbally as
follows:

"This is Investigator _____ of the U.S. Food and
Drug Administration speaking in the (state location) of (firm
name), (address), (city), (state), and (zip code). It is now
a.m./p.m. on (date). Present are (list individuals present
with title). This discussion is being recorded by both the
representative of (firm name) and by me. We are going to
discuss the inspectional findings of an inspection conducted
at this firm on (inclusive dates)."

At the close of the discussion and prior to leaving the firm,
the recording will be verbally identified as follows:

"This is Investigator _____ speaking. It is now
_____ a.m./p.m. on (date). This was a recording of the
discussion with management at the conclusion of an
inspection of (firm name and address) conducted on
(dates)."

If the recording covers a different situation, the identification
should be modified accordingly. If the representative of the

firm refuses permission to record the discussion, continue with your discussion and report the facts in your EIR.

The tape cassette must be identified with the firm name, date of the inspection, and investigator's name. Districts have the option of transcribing the tape and making the transcription an exhibit for the EIR. However, the tape itself must be made a permanent part of the EIR as an exhibit.

5.3.6 - Responsible Individuals

The identification of those responsible for violations is a critical part of the inspection, and as important as determining and documenting the violations themselves. Responsibility must be determined to identify those persons to hold accountable for violations, and with whom the agency must deal to seek lasting corrections.

Document and fully report individual responsibility whenever;

1. It is required by the assignment,

2. Inspectional findings suggest the possibility of regulatory action, or

3. Background information suggests the possibility of regulatory action.

Under the Medical Device Quality System regulation (21 CFR 820.20), if the management at the firm is not exercising the controls required by the regulation, the deviations may be cited on your FDA 483.

5.3.6.1 - Discussion on Duty, Power, Responsibility

Duty - An obligation required by one's position; a moral or legal obligation.

Power - Possession of the right or ability to wield force or influence to produce an effect.

Responsibility - An individual who has the duty and power to act is a responsible person.

Three key points to consider are:

1. Who had the duty and power to detect the violation?

2. Who had the duty and power to prevent the violation?

3. Who had the duty and power to correct the violation?

5.3.6.2 - Inspection Techniques How to Document Responsibility

Always determine and report the full legal name and title of persons interviewed, who supplied relevant facts and the name/title/address of top management officials to whom FDA correspondence should be directed.

Obtain the correct name and correct title of all corporate officers or company officials. Obtain pertinent educational and experience backgrounds, and the duties and powers of the officers and employees in key managerial, production, control, and sanitation positions. Ascertain the experience and training of supervisory personnel, in terms that will describe their qualifications to carry out their responsibilities.

There are numerous ways to establish and document responsibility. Evidence may be obtained during interviews and record review specifically intended to determine responsibility. Cover and report items such as:

1. Organizational charts,

2. Statements by individuals admitting their responsibility or attributing responsibility to others,

3. Company publications, letters, memos and instructions to employees, and

4. The presence or absence of individuals in specific areas at specific, significant times, and their observed activities directing, approving, etc.

In order to establish relationships between violative conditions and responsible individuals, the following types of information, would be useful:

1. Who knew of conditions?

2. Who should have known of the conditions because of their specific or overall duties and positions?

3. Who had the duty and power to prevent or detect the conditions, or to see they were prevented or detected?

4. Who had the duty and power to correct the conditions, or to see they were corrected? What was done after person(s) learned of the conditions? Upon whose authority and instructions (be specific)?

5. What orders were issued (When, by whom, to whom, on whose authority and instructions)?

6. What follow-up was done to see if orders were carried out (when; by whom; on whose authority and instructions)?

7. Who decided corrections were or were not complete and satisfactory?

8. What funding, new equipment, new procedures were requested, authorized or denied in relation to the conditions; who made the requests, authorizations, or denials.

Duties and power related to general operations should be established to supplement the specific relationships to violations. Examples of operational decisions that indicate responsibility are:

1. What processing equipment to buy.

2. What raw materials to purchase.

3. What products to produce and what procedures to follow in production?

4. Production schedules - how much to produce, what to make, when to stop or alter production?

5. What production controls to be used?

6. What standards are set for products, raw materials, processes?

7. How to correct or prevent adverse conditions; how much to spend and whom to hire to correct or prevent adverse conditions; when to clean up?

8. How products will be labeled; what products to ship; label approval?

9. When to reject raw materials or products; when to initiate a recall; acceptable quality levels for products?

10. When to hire or fire personnel?

11. Who will accept FDA 482, Notice of Inspection; refuses inspection; accept Inspectional Observations, FDA 483?

12. Who designed and implemented the quality assurance plan; who receives reports of Q.A.; who acts or should act upon the reports?

13. Who is responsible for auditing other facilities, contractors, vendors, GLP sites, etc.?

14. In the firm's business relationships, who signs major contracts, purchase orders, etc?

In some circumstances, documenting of individual responsibility requires investigative techniques that lead to sources outside the firm. These sources may include contractors, consultants, pest control or sanitation services, local health officials and others. Copies of documents between the firm and outside parties may help establish responsibilities. Do not overlook state officials as another possible source of information in selected cases.

During the course of the inspection you may observe persons who hold responsible positions and/or influence in the firm whose abilities or judgment may be affected by an obvious infirmity, handicap, or disability. If it is obvious the infirmity adversely affects the person's responsibilities or duties that are under FDA oversight, describe in your EIR the extent of the infirmity and how it relates to the purported problem or adverse condition.

5.3.7 - Guarantees and Labeling Agreements

Review the Code of Federal Regulations, 21 CFR 7.12, 7.13, 101.100(d), 201.150, and 701.9, for information concerning guarantees and labeling agreements.

5.3.7.1 - Guaranty

Certain exemptions from the criminal provisions of the FD&C Act are provided where a valid guarantee exists as specified in Section 303(c) of the FD&C Act [21 U.S.C. 333 (c)]. Obtain a copy of any Food and Drug guarantee, which the firm claims to use relating to a violation noted during your inspection. No person may rely upon any guaranty unless he has acted merely as a conduit through which the merchandise reached the consumer.

5.3.7.2 - Labeling Agreement

Products regulated by FDA are normally expected to be completely labeled when introduced into or while in interstate commerce. Under certain conditions exemptions are allowed when such articles are, in accordance with trade practices, to be processed, labeled, or repacked in substantial quantity at an establishment other than where originally processed or packed. Sections 405, 503(a) and 603 of the FD&C Act [21 U.S.C. 345, 353(a), and 363] also provide exemptions from complete labeling for products.

5.3.7.3 - Exemption Requirements

To enjoy this exemption, the shipment must meet one of the following:

1. The shipper must operate the establishment where the article is to be processed, labeled or repacked; or

2. If the shipper is not the operator of the establishment, he must first obtain from the owner a written agreement signed by and containing the post office addresses of such persons and such operator and containing such specifications for the processing, labeling or repacking of such articles as will insure that such article will not be adulterated or misbranded within the meaning of the Act, upon completion of the processing, labeling or repacking.

Submit copies and dates of labeling agreements where unlabeled articles are shipped in interstate commerce.

5.3.8 - Records Obtained

Many types of inspections and investigations require collection of copies of records to document evidence of deviations. In some cases, this may involve voluminous copies of Good Manufacturing Practice (GMP) records, commitments made in the Pre-Approval process, adherence to the requirements of the Low Acid Canned Food regulations or other areas. Copies of records are also obtained to document interstate commerce, product labeling and promotion, and to identify the party or parties responsible for a variety of actions. All documents become part of the government's case should it go to litigation.

Normally, during litigation proceedings, the best evidence rule prevails in court, whereby the copy of the record in the custody of the government can be authenticated, if the original record is not produced by the custodian of the record.

It is imperative the government witness [usually the collector of the record(s)] be able to testify where, when and from whom the copies were obtained, and that the copy is a true copy of the source document, based on their review of the source document.

5.3.8.1 - Identification of Records

Articles used as evidence in court cases must be marked to assure positive identification. This includes all records as noted in IOM 5.3.8, and any others for evidence in administrative or judiciary proceedings. When identifying and filing records, you must assure the record is complete and no identification method or filing mechanism covers, defaces or obliterates any data on the record/document.

It is imperative you identify the records used as evidence so you can later testify the documents entered as evidence are the very ones you obtained. See IOM 5.3.8.2. You should always review source documents to assure the records you obtained are an accurate representation (copy) of the source document. Record in your Regulatory Notes the when, where, and from who copies are obtained so you can properly prepare for testimony as needed.

5.3.8.2 - Identifying Original Paper Records

NOTE: Policy Changes - In keeping with other regulatory and enforcement agencies' policies, the mandatory identification of the original or source document copied during an inspection or investigation is no longer routinely required. IOM 4.5.2.5 covers identification of records collected and submitted as part of a sample collection.

When you collect an Official or Documentary (or "DOC") Sample, each page of the copied records will become part of the collection report and should be identified as noted in this section and as in IOM 4.4.5. This includes records of interstate commerce, manufacturing deviations, label and labeling violations, or any other record copied which may become "evidence.

While it is no longer routinely required for you to identify the original or source record(s), you must verify the copy of the record(s) you received is an accurate reproduction of the original or source record(s). You must be able to testify your copy is an exact duplicate of the original or source record. You should record in your regulatory notes you authenticated copies of records you obtain so you can provide this testimony during any trial proceedings.

To ensure you are able to positively identify the specific copies you received during your inspection or investigation and to avoid any filing mix-up, you must identify the copies you obtained. This identification will cover records submitted in support of the inspection or investigation, and include all those submitted whether it is an Establishment Inspection Report (EIR) or a narrative memorandum.

You should identify records/exhibits submitted with an EIR using at least the Exhibit number, firm name, date(s) of the inspection, and your initials. This should be done in such a way that you will be able to clearly identify the copy of specific record(s) you obtained. If some type of label is used, it must be permanently applied so any removal will be obvious. Records submitted with a Collection Report will be similarly identified with the sample number, date of collection, but with your handwritten initials. Records submitted with a memorandum will include a phrase or firm or subject name to tie them to the investigation, the date(s) of the investigation and your initials.

There are occasions when a single record may include hundreds of sheets of bound paper. Abbreviated methods of identification may be used for bound documents by fully identifying the first and last few pages. In some cases, firm's clearly mark each page with the sequential and total pages number (e.g., page 6 of 10, 7 of 10, etc.) and this allows you to fully mark only a few pages in the beginning and end of the exhibit.

All pages must be identifiable if not in bound documents. One example of a shortened method of identifying individual exhibits containing a large number of pages (usually more than 25) is to fully identify the first few and last few pages with at least the exhibit number, date and your initials. Then identify the remaining pages with the page number of the total page numbers, and your initials, e.g., "5 of 95 SHR". This may not be acceptable if you have more than one exhibit consisting of exactly 95 pages.

Whatever method is used, you must assure the document is complete and is always identifiable. This is so you can testify as to the "where", "when" and "from whom" the copies were obtained, and that the copy is a true copy of the source document based on your review of the source document. The identification method should allow any reviewer to determine if the document is complete or pages or parts are missing.

5.3.8.3 - Filmed or Electronic Records

When attempting to obtain records, you may find they are stored on microfilm, microfiche, or some form of a computerized management information system as electronic records.

5.3.8.3.1 - Microfilm/Microfiche and Electronic Information

You may encounter records stored on microfilm/microfiche or as electronic records on a computer system. Hard copy records obtained during the course of the inspection from these sources are handled the same as any hard copied records following procedures outline in IOM 5.3.8, 5.3.7.1 and 5.3.8.2.

NOTE: *See CPG Section 130.400 for Agency Policy concerning microfilm and/or microfiche records. 21 CFR Part 11 contains information concerning Electronic Records and Electronic Signatures and may be of value to you.*

5.3.8.3.2 - Electronic Information Received on CD-R, or other Electronic Storage Media

You may obtain electronic information, databases, or summary data from a firm's databases during an establishment inspection. The methods used must maintain the integrity of the electronic data and prevent unauthorized changes. Do not personally access a firm's electronic records, databases, or source/raw data during the course of an inspection.

When it is necessary to access a firm's data during an inspection:

1. Oversee the firm's personnel accessing their system and have them answer your questions.

2. Request the firm run queries specific to the information of interest.

3. Have the firm generate reports/data to be copied to a CD or other electronic storage media, which you can subsequently analyze, or have the data printed in hardcopy.

Electronic data, such as blood bank databases, drug production records, medical device complaints, service records, returned products and other records are often dynamic data files with real time updating. Information from these files is generally provided at the time of the inspection. Your request may require the firm to develop one or more custom queries to provide the requested information. You must assume the query logic is not validated and take appropriate action to ensure the data is accurate and no data has been accidentally omitted due to a programming logic error occurring at the firm.

When appropriate, a copy of electronic data can be obtained on one or more CD-R, or other electronic storage media. If you provide the diskettes to the firm, use only new, previously unused and preformatted diskettes. An additional safeguard is to request the firm reformat the disk on their own computer to assure it is usable and "clean".

Any request for electronic information on a CD-R, or other electronic storage media must be made with a computer application in mind and the data obtained must be useful. Request for electronic information should be in a format compatible with software applications knowledgeable to you and available from the Agency. Converting files into different file formats is difficult and should not be attempted without the necessary knowledge and availability of conversion type programs where applicable. If help is needed for file conversion, assistance may be available within the district, region or from DFI HFC-130.

Any CD-R or other electronic storage media containing electronic information received during the course of an inspection should be considered and handled as master CD-Rs. The firm may or may not retain a copy of the information provided during the course of an inspection. Ask the individual providing the copy(s) to provide actual CD-R or other electronic storage media labeling information, such as filename(s), date and other information to facilitate their later identification of the CD-R or other electronic storage media and the data provided on the CD-R or other electronic storage media. The name of the appropriate software and version used to ensure readability of the information should also be maintained with the copy of the electronic information.

You should perform a virus scan of the master CD-R or other electronic storage media according to Agency requirements. Each master diskette should be write-protected, labeled and identified as you would any hard copy document.

There are no guarantees the files provided on CD-R or other electronic storage media will be useable data. It is your responsibility to make a working copy of each master CD-R or other electronic storage media. Before making any working copies from the master CD-R or other electronic storage media, confirmation should be made that the write-protection has been activated on each master diskette. You will need to use a computer to view the copied files and verify each file contains the information requested and the information is useable to you. Some electronic data files may be too large to open from a CD-R or other electronic storage media and must be loaded on a hard disk before opening. If this is the case, the file should be put on a subdirectory before opening and viewing.

As a general practice, any findings developed from electronic information provided by the firm should be requested in a hard copy format. The hard copy provided by the firm should then be used as an exhibit to support the investigator's observation. This will preclude or limit any errors that may have occurred from the investigator querying of the electronic information.

The master CD-R, diskettes or other electronic storage media, should be secured to assure the integrity of the data when used in a subsequent enforcement action. Identify the master CD-R as an exhibit, write-protect diskettes, and place in a suitable container, e.g., FDA-525, and officially seal. Mark the FDA-525 or other container as containing diskettes and to "Protect from magnetic fields." The diskette(s) should be stored as part of the exhibits with the original EIR. See IOM 5.10.5.1.

5.3.8.4 - Requesting and Working with Computerized Complaint and Failure Data

The auditing of FDA regulated firms has found that an increasing number of firms are developing and maintaining computerized complaint and failure data to meet GMP record requirements. Records, hardcopy and electronic, are becoming increasingly voluminous. The auditing of information contained in computerized databases is generally most effectively accomplished with the use of a computer.

Computer auditing of computerized complaints and failure data may require the transfer of electronic data to CD-R or other electronic storage media for you to use in your computer. You should use a computer and application software familiar to you to query information obtained in electronic format. You should not use the audited firm's equipment or personnel to perform repetitive queries or manipulation of the audited firm's own computerized data.

5.3.8.4.1 - Computerized Complaint and Failure Data

Requesting and obtaining electronic data on CD-R or other electronic storage media is becoming more common during the course of routine inspections. Providing computerized data on electronic media is advantageous to both you and the firm and can result in shorter inspection time. These types of databases contain large numbers of records, which can be easily and quickly queried if they are in electronic format. Inspection time would be lengthened if all such information was only provided in hardcopy format. It may result in you reentering all of the hardcopy data into a new database or reviewing volumes of documents. Be aware if the firm should generate custom software to provide requested electronic records, it would be difficult for you to validate or verify the firm's algorithm used to extract the requested data and ensure that records were not accidentally or deliberately omitted due to programming logic errors, data entry errors, etc.

5.3.8.4.2 - Requesting Computerized Data

Before requesting a copy of computerized data, you should determine several things including information about the size and contents of the database, the program used by the firm, and the program you will use, among others. The following steps are useful in preparing for an electronic record request.

1. Determine the firm's application program used to maintain the data of interest. This may be in a DOS compatible application program such as Access, Excel, Dbase, Paradox, Lotus 123 or others. It is best to obtain data files in a format compatible with application programs you will be using. Large data files with record counts in excess of 10,000 records are best converted to file formats that can be used by programs designed to handle such large databases. There are spreadsheet record limits in some commercial programs that would not allow these application programs to handle much over 5,000 records. Check the program you plan to use to ensure it can handle the file size you will be using.

2. Most large and real-time data files reside in mainframe or network systems requiring programming and downloading to a PC using an [Structured Query Language (SQL)] SQL format. Although data may be captured and downloaded in an SQL format, not all spreadsheet or database application software can load an SQL file. In addition, it may be difficult or impossible to manipulate data in that format. Problems can also be encountered downloading data from Apple computers to an IBM format. Successful conversions are possible if the firm selects the proper conversion format or you have conversion software designed to convert from an Apple to an IBM platform.

3. You may need to request an ASCII (American Standard Code for Information Interchange) text/flat file format. ASCII format is an industry standard, which assigns a unique code to every printable, keyboard, and screen character. An ASCII file should be stripped of all non [-] standard codes that are used by specific application programs for fonts, underlining, tabs, etc. The ASCII text file can be imported by all application programs, and once imported, can be restructured for the specific application program. ASCII delimited is the format of choice, with ASCII fixed length as an alternative. Care must be exercised in specifying a hard carriage return at the end of each line to be DOS compatible, or additional conversion may be necessary before the file is useable.

4. You should determine what fields of information are routinely captured by the firm. This can be accomplished by requesting a printout of the data structure of the data file or observing the inputting of data at a computer terminal or workstation. It is common for databases to contain numbers or other coded information requiring translations from look up tables to give meaningful text. You should determine if information fields contain coded data, and if so, a code breakdown should be obtained. Information about code breakdowns should be located in the SOPs for that computerized system. Also be aware in relational databases, there may be linking data fields that exist in other tables that should also be considered in the overall data request.

5. If the files are too large to fit on a disk, file compression must be used. If possible, ask that the firm prepare the data in a compression format that is self-extracting. Self-extracting files are executable files and should be virus scanned before and after executing. All CD-R, diskettes or other electronic storage media should be scanned prior to being used on any FDA computer. Whatever compression utility is used, make sure you have the software to manipulate the files as needed.

6. You should always get the total record count of the data file provided by the firm. This count should be verified any time the file is loaded, converted, manipulated, or queried.

5.3.8.4.3 - Identification and Security of CD-R, Diskettes or Other Electronic Storage Media

You should follow these steps to ensure proper identification and security of CD-R or other electronic storage media:

1. Label each CD-R or other electronic storage media

   a. Firm name

   b. Date and your initials

   c. Initials by a representative of the firm (optional) If you provide the diskettes to be used, use only new and preformatted diskettes from an unopened box.

        d. The name of the appropriate software and version to ensure readability of the information

2. Make a working copy of CD-R or other electronic storage media

        a. Write protect the original diskette

        b. Virus scan the original diskette

        c. Copy the original CD-R or other electronic storage media

The original CD-R or other electronic storage media should not be used for manipulating data so as to maintain the integrity of the CD-R or other electronic storage media and data. NOTE: If a virus is detected, do not remove the virus from the source diskette provided by the firm. This may become evidence if it is suspected that the firm intentionally transferred the virus. Attempt to obtain another, uninfected copy of the data file from the firm.

Create a subdirectory on the computer hard drive:

1. Transfer data from the virus-free, working copy of the CD-R or other electronic storage media to your hard drive.

2. Virus scan any decompressed files before and after decompression. (Some virus scan software will scan compressed files but it is safer to scan all foreign files

3. You have now transferred confidential information to the hard drive and that information must be protected.

4. Upon completion of the use of the data, the file must be deleted and totally overwritten with a utility to wipe the data from the hard drive. A delete file operation is not adequate to totally remove the data from the hard drive.

5. Do not leave confidential files in any shared directories or e-mail.

5.3.8.4.4 - Data Integrity of Records Provided by Firm

Many manufacturers are using computers to store records concerning complaints, failure data, returned goods, servicing, testing results and others. Record traceability and data integrity are always concerns when you copy or use computerized data.

1. It is difficult to determine what records are to be designated as originals or copies of original records. It is important, when obtaining hardcopy or copy of computerized data, for you to capture some method of dating. The date of an electronic file can be captured by recording the date and time from a file listing in DOS or with File Manager in Windows. This may not always be possible, but some attempt should be made to date and time stamp electronic data.

2. Requests for most information from manufacturers will require the use of some custom software routine to generate the Investigator's requested information. Any data generated at the request of an Investigator should always be considered custom data. The firm will seldom validate or verify software routines used to generate data in response to your request. You should request a copy of any software program or scripts used to generate the computerized data provided. The request for the software program is not a request for a copy of the application program but a request for the special commands or programs created within the application program for the querying and extraction of data into a new data file. You should review the command structure to ensure it includes all data related to your request.

5.3.8.4.5 - Electronic Information for Official Documentation

During your use of queried data, if you find a violative situation, you should request the firm prepare a hardcopy report of the specific data that depicts the situation. (Do not request an entire copy of the data base and do not rely on the digital database or your extractions from the data to serve as official documentation.) Any records of interest, such as complaints, failure information, etc., noted from querying the computerized data should be copied from original hardcopy documents to support the findings in the database. You should also maintain the procedures or commands you used to find the violative situations in the data base. Follow procedures in IOM 5.3.8.4.3 for maintaining and identifying original disks.

5.3.8.5 - Listing of Records

If management requests a list of the copies of records you obtain, prepare it in duplicate and leave the original with the firm. Many firms prepare duplicate copies of documents requested during our inspections. In the interests of conserving inspectional time, you may ask the firm to prepare the list of copies concurrently with the photocopying and you then verify the accuracy. Do not use form FDA-484, Receipt for Samples. Describe the circumstances in your report including the name and title of the individual to whom you gave the list. Submit the duplicate list with your report as an Exhibit.

5.3.8.6 - Patient and/or Consumer Identification on Records

During the course of many types of inspections and investigations you will review and collect records which specifically identify (by name) patients or consumers. Under most state Privacy Laws this information is confidential. Some firms we inspect may mistakenly believe this information is not releasable to the federal government. However, Federal laws preempt State laws; with few exceptions we are entitled to review and copy the complete record, including the identifying patient/consumer names. The Agency is then required to maintain the confidentiality of the records/files, as with any confidential record you collect. Any disclosure of the information contained in the record(s) can only be by Law, i.e., judge's order, disclosure, Congressional order, etc.

General, routine guidance is as follows:

1. For records copied as a result of injury or complaint investigation, where you obtain patient identification, the identification should remain intact and stored in the official FDA files. Frequently, medical releases must be obtained from a complainant, consumer or "next-of-kin". At least one or two extra should be obtained and stored in the files.

2. For methadone inspections, continue the Agency policy of deleting patient identification specific to the patient (name, SSN, Driver License #, etc.).

3. For any inspection/investigation involving a regulation required Informed Consent, such as clinical

investigations, IRBs, bioequivalence testing, etc., patient identification should remain intact and stored in the official FDA files.

4.   For most others, such as MQSA, plasmapheresis, blood donations, etc., only the patient initials and unique identifier supplied by the firm (such as donor number, donation number, etc.) need be routinely retained in the FDA files.

It is not uncommon for a firm to voluntarily purge the documents of the pertinent identifiers as they are copied. You must verify (by direct comparison to the original document) you received an accurate reproduction of the original, minus the agreed to purging, prior to accepting the copy.

As with any inspection there are times when the specific identifiers must be obtained, copied and retained, such as if/when further interview of the patient/consumer could be necessary. If in doubt, obtain the data. It is always easier to delete later than to return to obtain the information, especially in the few cases where questionable practices may result in the loss of the information.

All documents obtained containing confidential identifiers will be maintained as all documents obtained by FDA containing confidential information, i.e., in the official FDA files. Confidential identifiers may be flagged in the official FDA files for reference by reviewers to assure no confidential data are released under FOIA.

5.3.9 - Request for Sample Collection

There are times one district will request another district to collect surveillance or compliance samples for it. The requesting district should provide as much of the following information as is available on specific shipments, using the FACTS Create Assignment Screen. See IOM Exhibit 5-9.

The following fields must be completed in order to save the assignment: Requesting Organization, Priority, Subject, POC Name, Op Code, Accomp Org, Num of Ops, and PAC. When you create a sample collection assignment, which will require laboratory analysis, you should also create an assignment for the laboratory, using operation 41.

The screen is organized in sections.

5.3.9.1 - FACTS Assignment Section

The Assignment section has the following fields:

Compliance Number: Enter the Compliance Number if known. This will make it easier to tie all associated activities together if the District is considering a compliance action. You can generate a compliance number after completing the mandatory fields on the Maintain Inspection Results screen.

ORA reqd: This field only applies to assignments generated by Centers or other organizations outside of ORA. It will indicate whether or not ORA concurrence is required for the assignment.

ORA Cncrnc Num: This field is for the requesting organization (other than an ORA component) to indicate ORA concurrence for the assignment.

POC Name: This field Indicates the point of contact in the requesting organization for the assignment.

Priority: Choose High or Routine

Remarks: This is a free form field, which should briefly describe the assignment.

Reporting Method: Indicate how the other district should notify the contact of problems with or status of the assignment. For example: e-mail, phone, etc.

Requesting Organization: Enter your District Office, if you are requesting a sample from another district or other appropriate FACTS organization.

Requestor Completion Date: Enter the completion date desired, using the format, MM/DD/YYYY.

Subject: Enter a subject for the assignment. It may be helpful to create a subject others will recognize as related to a specific action, for example a firm or product name.

5.3.9.2 - FACTS Operations Section

The Operations section has the following fields:

Estmtd Hours: Enter the number of hours you believe the assignment should take. This is done to assist the collecting district in planning their work.

Estmtd Smpl Cost: Enter the estimated sample cost, if known.

Op Code: Enter the operation code for the assignment. If you are requesting a sample collection, it is 31.

Requester Remarks: Enter as many details about the sample collection as you can. Include: date of shipment, number and size of units or amount, codes, carrier (routing and freight bill number), invoice number, and name of responsible firm with date of inspection (if one occurred).

Rqstr Prty: Enter High or Routine. This will default to the same data entered in the Assignment section if it was prepared first.

Subject: This will default to the same data entered in the Subject field in the Assignment section if it was prepared first.

5.3.9.3 - FACTS Organizations Section

The Organizations section contains the following fields.

Accomp Org: Enter the District or other FACTS organization you are requesting collect the assignment. If you are completing the sample analysis assignment, be sure to enter a laboratory.

Num of Ops: Enter the number of sample collections or analyses you are requesting from the organization identified in the previous field.

Perf Org (Adhoc Work): If the performing organization is part of the accomplishing organization you are in, you may enter the performing organization here. If you are requesting the sample of another District, you will probably leave this blank.

The PACS and Products section of the form contains fields for entering the assignment PAC and Product code.

Enter the FEI number(s)/CFN(s) of the firm or firms from which the sample is to be collected in the Firms and Cross References section. See IOM 4.4.10.3.24.

5.3.10 - Post-Inspection Notification Letters

Issuance of Post-inspection notification letters have been discontinued in all program areas. See FMD 145.

5.4 - Food

5.4.1 - Food Inspections

Food plant inspections are conducted to evaluate the methods, facilities, and controls used in manufacturing, storage and distribution of foods.

See CFSAN Office of Compliance's intranet website for the most current guidance (e.g., compliance programs, field assignments, field guidance).

5.4.1.1 - Preparation and References

Before undertaking an inspection:

1. Review the district files of the firm to be inspected and acquaint yourself with the firm's history, related firms, trade marks, practices and products. The review will identify products difficult to manufacture, require special handling, special processes or techniques, and hours of operation, which is especially important in bacteriological inspections. Remove, for subsequent investigations and discussion with management, Complaint/Injury Reports, which are marked for follow-up during the next inspection. See IOM 5.2.8.

2. Become familiar with current programs relating to the particular food or industry involved and relevant DFI inspection guides. Become familiar with any applicable Compliance Policy Guide (CPG Chap 5).

3. Understand the nature of the assignment and whether it entails certain problems, e.g., Salmonella or other bacteriological aspects.

4. Review the FD&C Act Chapter IV - Food.

5. Review and become familiar with the appropriate parts of 21 CFR pertaining to foods, for example:

   a. 21 CFR Part 110 - GMP's on foods

   b. 21 CFR Parts 108 and 113 - Thermally Processed Low-Acid Foods Packaged in Hermetically Sealed Containers

c. 21 CFR Part 114 - Acidified Foods are of particular importance

d. 21 CFR Part 120 - HACCP Systems (covers Juice Processors)

e. 21 CFR Part 123 - Fish and Fishery Products

f. 21 CFR Part 129 - Processing and Bottling of Bottled Drinking Water

g. 21 CFR Part 130, et al - Food Standards

h. 21 CFR Part 1240 - Control of Communicable Disease

i. 21 CFR Part 1250 - Interstate Conveyance Sanitation

6. Review reference materials on food technology and other subjects available in the District Inspectional Reference Library.

7. If you are assigned to inspect food-service establishments under the FDA - Secret Service Agreement, you should use the most current copy of the "Food Code" and be standardized in its use. All Regional Food Service Specialists and most Interstate Travel Sanitation Specialists are standardized in use of the code.

8. Be familiar with the "Food Chemicals Codex". See IOM 5.4.4.3.

5.4.1.2 - Inspectional Authority

See IOM subchapter 2.2 for broader information on this topic.

Authority to Obtain Records and Information in LACF and Acidified Foods Plants:

FDA's regulation in 21 CFR 113 requires commercial processors of low-acid foods packaged in hermetically sealed containers to maintain complete records of processing, production and initial distribution. 21 CFR 114 requires the same of commercial processors of acidified foods. 21 CFR 108.25(g) and 21 CFR 108.35(h) provide that a commercial processor shall permit the inspection and copying of the records required by 21 CFR 113 and 21 CFR 114 by duly authorized employees of FDA. The demand for these records must be in writing on an FDA 482a, Demand for Records, signed by you and must identify the records demanded.

5.4.1.2.1 - Written Demand for Records

To obtain the records:

1. Prepare a FDA 482a, "Demand for Records", listing the records demanded. Describe the processing records to be reviewed and/or copied as accurately as you can, e.g., "All thermal process and production records mandated by 21 CFR 113 (or 114 if applicable) for the foods (state name of food) processed at this plant on (specific date or period of time)". If only a specific record is desired list it specifically as follows: e.g., "Fill Weight Records for #2 Filling Machine for the period 4-15-87 through 6-7-87."

2. Sign the form.

3. Issue the original to the same person to whom the FDA 482, "Notice of Inspection", was issued.

4. Submit the carbon copy with your EIR.

5.4.1.2.2 - Written Request for Information

21 CFR 108.35(c)(3)(ii) states commercial processors engaged in thermal processing of low-acid foods packaged in hermetically sealed containers shall provide FDA with any information concerning processes and procedures necessary by FDA to determine the adequacy of the process. 21 CFR 108.25(c)(3)(ii) requires the same of commercial processors of acidified foods. The information in this regulation is the data on which the processes are based. Many processors will not have this information and in fact 21 CFR 113.83 requires only that the person or organization establishing the process permanently retain all records covering all aspects of establishing the process. The processor should, however, have in his files a letter or other written documentation from a processing authority delineating the recommended scheduled process and associated critical factors.

You may encounter situations where you believe control of certain factors is critical to the process and there is no evidence to document these factors were considered when the process was established (e.g., a change in formulation which could effect consistency). It is appropriate to issue a written request for a letter or other written documentation from a processing authority, which delineates the recommended scheduled process and associated critical factors. This represents the processing authority's conclusions and should correlate with the filed process.

If you believe control of certain factors are critical to the process and are not delineated in the process authority's recommendation or the filed process, obtain all available information about the situation. Include the name of the person or organization who established the process and the specific practices of the firm. This information should be included in your report and forwarded by your District to the Center for Food Safety and Applied Nutrition, Division of Enforcement (HFS-605) for review, as soon as possible. If the process establishment data and information is deemed necessary by the center, they will either request it directly from the processor or will direct the district to request it. If requested to obtain the information:

1. Prepare a FDA 482b - Request for Information listing the specific information requested. Specify each product involved by food product name and form, container size and processing method.

2. Sign the form.

3. Issue the original to the same person to whom the FDA 482, "Notice of Inspection", was issued.

4. Submit the carbon copy with your EIR.

5.4.1.3 - Records Access Under BT Authority

The following guidance should be used to demand records under the Agency's Bioterrorism (BT) authority when the following conditions are met:

1. The Secretary has a reasonable belief that an article of food is adulterated and presents a threat of serious adverse health consequences or death to humans or animals.

2. The records are necessary to assist the Secretary in making such a determination. When these conditions are met, the following guidance should be followed: http://www.cfsan.fda.gov/~dms/secgui13.html.

FDA will not invoke this authority during routine inspections unless the requirements for record access under the BT Act are satisfied. Note: FDA will continue to request that food records be voluntarily provided by the owner, operator, or agent in charge in a variety of circumstances of routine inspections. The procedures and the guidance above will be followed only if records are requested under the Agency's BT Act authority.

Procedure: Upon the determination that a food presents a threat of serious adverse health consequences or death to humans or animals, and after concurrence by the individuals listed in the guidance above, a FDA 482c (Exhibit 5-10) will be issued to the most responsible individual.

5.4.1.4 - Food and Cosmetic Defense Inspectional Activities

Food and cosmetics security inspectional activities should be conducted during all routine food and cosmetics safety inspections. During the normal course of the inspection be alert to opportunities for improvement or enhancement of the firm's food and cosmetics security preventive measures, as compared to those recommended in the guidance documents described below. You should not perform a comprehensive food and cosmetics security audit of the firm or conduct an extensive interview of management or employees in an attempt to determine the level of adoption of preventive measures listed in the guidance. The goal is to facilitate an exchange of information to heighten awareness on the subject of food and cosmetics security.

5.4.1.4.1 - Food and Cosmetic Security

Inspectional activities relative to food and cosmetic security for routine food and cosmetic establishment inspections should include:

1. Discussion with firm management of relevant FDA guidance documents including:

a. Food Producers, Processors, and Transporters: Food Security Preventive Measures Guidance

b. Importers and Filers: Food Security Preventive Measures Guidance

c. Cosmetics Processors and Transporters: Cosmetics Security Preventive Measures Guidance

d. Retail Food Stores and Food Service Establishments: Food Security Preventive Measures Guidance

e. Dairy Farms, Bulk Milk Transporters, Bulk Milk Transfer Stations, and Fluid Milk Processors: Food Security Preventive Measures Guidance.

These documents should be used as references during inspections, as appropriate. Copies may be obtained at: http://www.fda.gov/oc/factsheets/foodsecurity.html. If firm management does not already have a copy of the relevant guidance documents provide them with hard copies or information on how to obtain the guidance from FDA's web site.

2. Identification of opportunities for improvement or enhancement of the firm's food and cosmetic security preventive measures, as compared to those recommended in the guidance documents, and encouragement of management to make such improvements or enhancements to their security system.

Keep in mind that: the guidance does not represent mandatory conditions or practices; some of the recommended food and cosmetics security preventive measures may not be appropriate or practical to the specific operation; and other means of achieving the goals of the preventive measures listed in the guidance may be more suitable for the specific operation than those cited as examples. The important message for management is to consider the goals of the food and cosmetics security preventive measures; evaluate the goals relative to the specifics of their operation; and address those that are relevant to the extent practical.

Food and cosmetics security observations should not be listed on form FDA-483, Inspectional Observations, unless they likewise constitute deviations from Current Good Manufacturing Practice. Security discussions should be handled discretely and should only involve management of the firm.

The fact that the discussion took place and, if applicable, that a copy of the guidance document(s) was provided should be recorded in the Summary section of the EIR. For example, under a section heading titled "Food and Cosmetics Security" you should only state, "A copy of the Food and Cosmetics Security Guidance documents were provided to and food and cosmetics security issues were discussed with (name of firm official)." The details of inspectional findings regarding security should NOT be recorded. You should also minimize the quantity and detail of notes taken relative to the firm's food and cosmetics security program, taking only those needed to serve as a "memory jog" during the discussion with management.

5.4.1.4.2 - Reconciliation Examinations

During routine food and cosmetic inspections, conduct one reconciliation examination during each food and cosmetic establishment inspection. The examinations are to be conducted on raw materials used in the manufacture of foods or cosmetics, or finished products received by the firm for further distribution. Preference should be given to products of foreign origin. Where possible, these examinations should be performed on products as they are received by the firm.

Consult the factory jacket for any information on special conditions in the facility that may affect selection of personal protective equipment; consult your supervisor for any recommendations on personal protective equipment; and have available all necessary personal protective equipment to conduct the activity.

As Part of an Import Field Examination and Entry Review - See IOM 6.3.1 and 6.4.3. For imported food and cosmetics, a reconciliation examination should be conducted:

Per Part A [IOM 5.4.1.4.3] during all routine import field exams. You should only report time under the Counter Terrorism PAC at the direction of your supervisor or if there is a for cause assignment.

1. In instances where review of entry information raises suspicion (resulting in a detailed reconciliation exam per Part B [IOM 5.4.1.4.4]).

2. A detailed reconciliation exam should be conducted when there are anomalies in entry declaration information. These may include new, unusual, or unfamiliar commodities, manufacturers, importers; suspicious trans-shipments; or credibility issues such as those between the product and declared country of origin.

   If anomalies are found, entry documents should be requested and reviewed for discrepancies between the information declared through electronic filer submissions and that found in entry documents. Entry documents may include invoices, bills of lading, export certifications, and other relevant documents obtained from the importer, filer, or manufacturer/processor of the product. Fields in which discrepancies are found that may raise concern include country of origin, manufacturer, product description, product code, and quantity.

   Avoid duplication of examination of the same foreign manufacturer, unless a prior reconciliation examination disclosed an unexplained discrepancy.

Follow guidance in IOM 5.4.1.4.3 to IOM 5.4.1.4.4 below for domestic and import reconciliation exams.

5.4.1.4.3 - Reconciliation Examination Guidance Part A

Reconciliation examinations are performed to ensure that:

1. The food or cosmetic is what it purports to be

2. There are not unexplained differences in the quantity of product ordered, shipped, and received, and

3. There are no signs of tampering or counterfeiting.

Before initiating the exam make a general assessment of the appearance of the lot. Look for packaging that: appears to have been opened and resealed; appears wet, stained, punctured, or powdered. Also be alert to abnormal chemical odors. If any of these conditions are detected stop the exam and contact your supervisor for guidance. If the lot appears normal proceed with the examination. To the extent possible the exam should be performed in a well-ventilated, well-lit area.

Determine, to the extent possible, whether:

1. The actual goods in a lot are the same as those that are declared in the shipping documents

2. There is consistency in the manufacturer declared on the product labeling, bulk product packaging, and shipping documents; and

3. There is no (unexplainable) inconsistency in actual quantity of goods in the lot, and the quantity ordered and declared in the shipping documents.

If no unexplained inconsistencies are detected, no further action is indicated.

If unexplainable inconsistencies are detected, document the occurrence, including photographs of the labeling and packaging, and an accurate count of the lot. Contact your supervisor, who should, in the case of imported products, contact the U.S. Customs and Border Protection for appropriate action. If the examination discloses evidence that inaccurate product identification data was submitted to the OASIS entry screening system, the District should evaluate the need for follow-up with a compliance filer evaluation and consider providing the information to the U.S. Customs and Border Protection for appropriate action.

In addition, if unexplained inconsistencies are detected, follow part B [IOM 5.4.1.4.4] of this guidance while conducting a detailed reconciliation exam.

5.4.1.4.4 - Reconciliation Examination Guidance Part B

Open the shipping packaging of a quantity of product approximating the square root of the number of shipping cartons/packages in the lot, and examine the contents. Look for the following:

1. Product identity on the package that does not match the identity declared on the shipping documents

2. Mixed product sizes within a carton or within the lot;

3. Product sizes that do not match the sizes declared on the shipping documents

4. Differences in product configuration or package type (e.g. plastic containers mixed with glass jars or aluminum or steel cans)

5. Easily apparent variations in weight

6. Product labels that display crude, unprofessional, or inconsistent styles of print, color or use of language

7. Unusual placement of labels (e.g. off-center)

8. Variations in lot coding ink color, appearance of embossing, or format (e.g., two line vs. three line, use of letters, numbers and symbols). unusually excessive use of a single code in a very large lot

9. Differences between the actual can codes in the lot and those listed on the shipping documents

10. The existence of a tamper-evident notice on the labeling when the packaging does not contain a tamper-evident feature

11. Product that is beyond its expiration date

12. Inconsistencies in expiration dates within a lot

If no unexplainable discrepancies are noted select at least 1 package at random from the entire shipment and examine their contents. For those products that the contents are visible through the package it is not necessary to open the package. For other products, open the package and examine and field destroy the contents. Look for the following:

1. Differences between the product and that which is declared on the label

2. Color differences in the product between containers of the same lot

3. Style differences in the product between containers of the same lot or between the actual product and the label and document declaration (e.g., sliced vs. whole, colorless noodles vs. egg noodles)

4. Readily detectable abnormal odors (e.g. strong decomposition, bitter almond, petroleum odor, garlic, chlorine, sulfur). Note: specific sensory examination is not expected.

Verification that the product is consistent with the product ordered may require that you obtain information from the owner of the goods, importer, filer, or custom house broker. Review of the following types of documentation may be necessary to accomplish the above instructions, to the extent that they are available: authentic label supplied by the owner of the goods, importer, filer, or custom house broker; purchase order; invoice; shipping records (bill of lading, weigh bill, manifest). Depending on the findings of the exam and record review, you may wish to request that the importer assist in an evaluation of the authenticity of the product, based on the importer's experience with the product.

Every effort should be made to document any discrepancies through use of photographs, and additional records that may be available from the filer, importer, owner, or customs house broker.

5.4.1.4.5 - Special Safety Precautions

See IOM Subchapters 1.5 Safety, subsections including 1.5.1.1 thru 1.5.1.4, and Section 1.5.3 on sampling hazards.

When performing an establishment inspection or reconciliation examination, follow these instructions:

1. If there are no signs of tampering or counterfeiting, use level I protection, which consists of: work gloves; coveralls; work boots; and in a dusty situation, a dust mask.

2. If there are signs of tampering or counterfeiting, use level II protection and consult your supervisor for any additional safety precautions needed. Level II protection consists of: work gloves worn over surgical gloves; full face respirator with appropriate cartridges; disposable coveralls; and work boots.

5.4.1.5 - Food Registration

The Public Health Security and Bioterrorism Preparedness and Response Act of 2002 (the Bioterrorism Act) requires most domestic and foreign facilities that manufacture/process, pack, or hold food for human or animal consumption in the United States to register with FDA by December 12, 2003. The Bioterrorism Act covers both interstate and intrastate firms. FDA published a final rule on October 3, 2005 (70 FR 57505) to implement this requirement. The regulations are codified at 21 CFR 1 Subpart H - Registration of Food Facilities. The agency has also issued a guidance to industry document, "Questions and Answers Regarding Registration of Food Facilities." Facilities may register electronically at http://www.access.fda.gov, by mail or fax, or by CD-ROM for multiple submissions. Registrations will be maintained in the FDA Unified Registration and Listing System (FURLS). Facilities are not considered to be registered until their information is entered into FURLS.

The owner, operator, or agent in charge of a facility must register the name and address of each facility at which, and all trade names under which, the registrant conducts business. If applicable, they must register the same information for their parent companies. Foreign facilities must also provide information about their U.S. Agent.

The purpose of registration is to provide sufficient and reliable information about food facilities. When used with the detention, recordkeeping, and prior notice provisions of the Bioterrorism Act, registration will help to provide information on the origin and distribution of food and feed products to allow for detection and quick reaction to real and potential threats to these products. In the event of a potential threat or an outbreak of foodborne illness, such information will help FDA and other authorities to notify food facility representatives and investigate the event, source, and/or cause of the outbreak. Also, it will enable FDA to notify quickly the facilities that might be affected by the outbreak.

For both domestic and foreign facilities, the Bioterrorism Act makes failure to register a prohibited act. In addition, food from an unregistered foreign facility may be held at the port of entry. Failure to register under the BT Act, does not make the food product(s) violative.

FDA estimates that the total number of food facilities that must register could exceed 400,000, including both domestic and foreign facilities.

5.4.1.5.1 - Facilities Exempted from Registration

The Bioterrorism Act, as implemented by the interim final rule for registration of food facilities exempts the following from registration:

1. A foreign facility, if food from such facility undergoes further manufacturing/processing (including packaging) by another facility outside the U.S. (Note: A facility is not exempt under this provision if the further manufacturing/processing (including packaging) conducted by the subsequent facility consists of adding labeling or any similar activity of a de minimis nature. The facility conducting the de minimis activity also must register;

2. Farms that are devoted to the growing and harvesting of crops, the raising of animals (including seafood), or both. Washing, trimming of outer leaves of, and cooling produce are considered part of harvesting. The term "farm" includes:

    a. Facilities that pack or hold food, provided that all food used in such activities is grown, raised, or consumed on that farm or another farm under the same ownership; and

    b. Facilities that manufacture/process food, provided that all food used in such activities is consumed on that farm or another farm under the same ownership.

3. Retail food establishments whose sales to consumers exceed their sales to non-consumers (businesses are considered non-consumers);

4. Restaurants that prepare and serve food directly to consumers for immediate consumption;

5. Nonprofit food establishments in which food is prepared for, or served directly to, the consumer;

6. Fishing vessels, including those that not only harvest and transport fish but also engage in practices such as heading, eviscerating, or freezing intended solely to prepare fish for holding on board a harvest vessel. However, those fishing vessels that otherwise engage in processing fish are required to register. For the

purposes of this section, "processing" means handling, storing, preparing, shucking, changing into different market forms, manufacturing, preserving, packing, labeling, dockside unloading, holding, or heading, eviscerating, or freezing other than solely to prepare fish for holding on board a harvest vessel;

7. Facilities that are regulated exclusively, throughout the entire facility, by the U.S. Department of Agriculture under the Federal Meat Inspection Act (21 U.S.C. 601 et seq.), the Poultry Products Inspection Act (21 U.S.C. 451 et seq.), or the Egg Products Inspection Act (21 U.S.C. 1031 et seq.)

Other exemptions from registration in the interim final rule are based on the definition of food included within the scope of the registration regulation. Facilities that manufacture/process, pack, or hold food contact substances (including packaging materials) or pesticides are exempt from registration.

5.4.1.5.2 - Agency Website Link

More specific information regarding the Bioterrorism Act and food registration may be obtained at the following website: http://www.cfsan.fda.gov/~dms/fsbtact.html.

5.4.1.5.3 - Inspectional Guidance

See Compliance Policy Guide Sec. 110.300: Registration of Food Facilities Under the Public Health Security and Bioterrorism Preparedness and Response Act of 2002. During inspections of domestic and foreign facilities subject to the rule, make sure that firm management is aware of the registration requirements. Inform the firm's management that information regarding food security, the BT Act, facility registration, required and optional information, definitions, exemptions, and penalties for failure to register, etc., is available at the following website: http://www.cfsan.fda.gov/~dms/fsbtact.html. For facilities that are required to register, but have not yet done so, encourage electronic registration (see http://www.fda.gov/oc/bioterrorism/bioact.html), and refer them to a copy of the blank registration form (see http://www.cfsan.fda.gov/~furls/helpol.html) and the web site address (http://www.access.fda.gov) for electronic registration. Also encourage submission of the optional information on the form to assist and facilitate future communication with the facility as intended by the BT Act.

Document the registration status of the firm, and registration discussions with firm management, in the "Summary of Findings and Discussion with Management" sections of the EIR. Per CPG 110.300, do not report the FURLS Registration number. Observations about failure to register are NOT to be placed on the FDA 483.

5.4.1.6 - CFSAN Bio-research Monitoring

Bio-research monitoring (BIMO) assignments for foods will generally be issued by the Center for Food Safety and Applied Nutrition (CFSAN) (see IOM 5.5.6).

5.4.2 - Personnel

5.4.2.1 - Management

Follow the guidance described in IOM 5.3.6 when documenting individual responsibility including obtaining the full name and titles of the following individuals:

1. Owners, partners, or officers.

2. Other management officials or individuals supplying information.

3. Individuals to whom credentials were shown and FDA 482, Notice of Inspection, and other inspectional forms issued.

4. Individuals refusing to supply information or permit inspection.

Individuals with whom inspectional findings were discussed or recommendations made.

Regulations require plant management take all reasonable measures and precautions to assure control of communicable disease, employee cleanliness, appropriate training of key personnel, and compliance by all personnel with all requirements of 21 CFR 110.10, 113.10, and 114.10.

Determine if adequate supervision is provided for critical operations where violations are likely to occur if tasks are improperly performed.

5.4.2.2 - Employees

Improper employee habits may contribute to violative practices in an otherwise satisfactory plant. Observe the attitude and actions of employees during all phases of the inspection. Observe employees at their work stations and determine their duties or work functions. Note whether employees are neatly and cleanly dressed and whether they wear head coverings which properly cover their hair.

Determine if employees working with the product have obvious colds, or infected sores, cuts, etc. Under no circumstance should you swab a sore, touch or remove a bandage from an employee in an attempt to obtain bacteriological data. To do so is a violation of personal privacy, possibly hazardous to you and/or the employee, and usually provides little useful data.

Note whether employees eat while on duty.

Observe and record insanitary employee practices or actions showing employees handling or touching unsanitized or dirty surfaces and then contacting food products or direct food contact surfaces. Such practices might include employees spitting, handling garbage, placing their hands in or near their mouths, cleaning drains, handling dirty containers, etc. and then handling food product without washing and sanitizing their hands. Observe whether employees comply with plant rules such as, "No smoking", "Keep doors closed", "Wash hands before returning to work", etc. See IOM 5.4.7.2.2.

Be alert to employees handling insanitary objects, then quickly dipping their hands in sanitizing solutions without first washing them. Depending upon the amount and type of filth deposited on the hands during the handling of insanitary objects, such attempts at sanitizing are questionable at best. Sanitizers work most effectively on hands, which have been first cleaned by washing with soap and water.

Conversations with employees doing the work may provide information on both current and past objectionable practices, conditions and circumstances. These should be recorded in your notes.

Where appropriate, determine employee education and
training. Also determine type, duration, and adequacy of
firm's training programs, if any, to prepare employees for
their positions and to maintain their skills. See IOM
5.10.4.1.

5.4.3 - Plants and Grounds

Observe the general nature of the neighborhood in which
the firm is located. Environmental factors such as proximity
to swamps, rivers, wharves, city dumps, etc., may
contribute to rodent, bird, insect or other sanitation
problems.

5.4.3.1 - Plant Construction, Design and Maintenance

Determine the approximate size and type of building
housing the firm and if suitable in size, construction, and
design to facilitate maintenance and sanitary operations.
Check placement of equipment, storage of materials,
lighting, ventilation, and placement of partitions and
screening to eliminate product contamination by bacteria,
birds, vermin, etc. Determine any construction defects or
other conditions such as broken windows, cracked floor
boards, sagging doors, etc. which may permit animal entry
or harborage.

Inspect toilet facilities for cleanliness, adequate supplies of
toilet paper, soap, towels, hot and cold water, and hand
washing signs. Check if hand washing facilities are hidden,
or if located where supervisory personnel can police hand
washing.

Determine who is responsible for buildings and grounds
maintenance. Many facilities such as docks, wharves, or
other premises are owned and maintained by other firms,
municipalities, or individuals for lease for manufacturing
operations. Determine who is legally responsible for repairs,
maintenance, rodent proofing, screening, etc. Evaluate the
firm's attitude toward maintenance and cleaning operations.

5.4.3.2 - Waste Disposal

Waste and garbage disposal poses a problem in all food
plants depending upon plant location and municipal
facilities available.

Check the effectiveness of waste disposal on the premises and ensure it does not cause violative conditions or contribute toward contamination of the finished products. Check for in-plant contamination of equipment and/or product, if its water is supplied from nearby streams, springs, lakes or wells.

Suspected dumping of sewage effluent into nearby streams, lakes, or bay waters near water intakes can be documented by color photographs and water soluble fluorescein sodium dye. Place approximately two ounces dye, which yields a yellowish red color, into the firm's waste system and/or toilets, as applicable, and flush the system. The discharge area of the effluent becomes readily visible by a yellowish-red color on the surface of the water as the dye reaches it. Color photographs should be taken.

Determine collecting or flushing methods used to remove waste from operating areas. If water is used, determine if it is recirculated and thus may contaminate equipment or materials.

Determine the disposition of waste materials that should not be used as human food such as rancid nuts, juice from decomposed tomatoes, etc.

Determine the disposition of waste, garbage, etc., which contain pesticide residues. Determine how this is segregated from waste material which contains no residues and which may be used for animal feed.

5.4.3.3 - Plant Services

If applicable, check steam generators for capacity and demand. Demand may reach or exceed the rated capacity, which could effect adequacy of the process. Check boiler water additives if steam comes in direct contact with foods.

Check central compressed air supply for effective removal of moisture (condensate) and oil. Determine if any undrained loops in the supply line exist where condensate can accumulate and become contaminated with foreign material or microorganisms.

5.4.4 - Raw Materials

List in a general way the nature of raw materials on hand. Itemize and describe those, which are unusual to you, or involved in a suspected violation (copy quantity of contents and ingredient statements, codes, name of manufacturer or distributor, etc.). Be alert for additives and preservatives. Evaluate the storage of materials. Determine the general storage pattern, stock rotation and general housekeeping. Materials should be stored so they are accessible for inspection. Thoroughly check ceilings, walls, ledges, and floors in raw material storage areas for evidence or rodent or insect infestation, water dripping or other adverse conditions.

5.4.4.1 - Handling Procedure

Determine if growing conditions relative to disease, insects, and weather are affecting the raw material. Check measures taken for protection against insect or rodent damage. Raw materials may be susceptible to decomposition, bruising or damage, e.g., soft vegetables and fruits delivered in truckload lots. Determine the holding times of materials subject to progressive decomposition.

5.4.4.2 - Condition

Evaluate the firm's acceptance examination and inspection practices including washing and disposition of rejected lots. Where indicated, examine rejected lots and collect appropriate samples and report consignees.

Determine the general acceptability of raw materials for their intended use and their effect on the finished product. Raw stocks of fruits or vegetables may contribute decomposed or filthy material to the finished product. Be alert for use of low quality or salvage raw materials. Check bags, bales, cases and other types of raw material containers to determine signs of abnormal conditions, indicating presence of filthy, putrid or decomposed items. Check any indication of gnawed or otherwise damaged containers, to ascertain if material is violative. Be alert to contamination of raw materials by infested or contaminated railroad cars or other carriers.

Document by photographs, exhibits or sketches any instances where insanitary storage or handling conditions exist.

5.4.4.3 - Food Chemicals Codex

Any substance used in foods must be food-grade quality. FDA regards the applicable specifications in the current edition of the publication "Food Chemicals Codex" as establishing food-grade unless FDA publishes other specifications in the Federal Register.

Determine whether firm is aware of this publication and whether or not they comply.

5.4.5 - Equipment and Utensils

By arriving before processing begins, you are able to evaluate conditions and practices not otherwise observable before plant start-up. This includes adequacy of clean-up, where and how equipment is stored while not in use, how hand sanitizing solutions and food batches are prepared and if personnel sanitize their hands and equipment before beginning work.

Dirty or improperly cleaned equipment and utensils may be the focal point for filth or bacterial contamination of the finished product. Examine all equipment for suitability and accessibility for cleaning. Determine if equipment is constructed or covered to protect contents from dust and environmental contamination. Open inspection ports to check inside only when this can be done safely. Notice whether inspection ports have been painted over or permanently sealed.

5.4.5.1 - Filtering Systems

Observe the firm's filtering systems and evaluate the cleaning methods (or replacement intervals of disposable filters) and schedules. Check types of filters used. There have been instances where firms have relied on household furnace type filters.

5.4.5.2 - Sanitation of Machinery

Check the sanitary condition of all machinery. Determine if equipment is cleaned prior to each use and the method of cleaning. If the firm rents or leases equipment on a short-term basis, report prior cleaning procedures. Equipment may have been used for pesticides, chemicals, drugs, etc., prior to being installed and could therefore be a source of cross-contamination.

5.4.5.3 - Conveyor Belt Conditions

Inspect conveyor belts for build-up of residual materials and pockets of residue in corners and under belts. Look in inspection ports and hard-to-reach places inside, around,

underneath, and behind equipment and machinery for
evidence of filth, insects, and/or rodent contamination.
Chutes and conveyor ducts may appear satisfactory, but a
rap on them with the heel of your hand or a rubber mallet
may dislodge static material, which can be examined. See
IOM 4.3.7.6 for procedure on taking In-line Sample Subs.

5.4.5.4 - Utensils

Determine how brushes, scrapers, brooms, and other items
used during processing or on product contact surfaces are
cleaned, sanitized and stored. Evaluate the effectiveness of
the practices observed.

5.4.5.5 - Mercury and Glass Contamination

Be alert for improper placement or inadequately protected
mercury switches, mercury thermometers, or electric bulbs.
Breakage of these could spray mercury and glass particles
onto materials or into processing machinery.

5.4.5.6 - UV Lamps

If firm is using ultra violet (UV) lamps for bacteria control,
check if it has and uses any method or meters to check the
strength of UV emissions. If so, obtain methods,
procedures, type equipment used, and schedule for
replacement of weak UV bulbs.

5.4.5.7 - Chlorine Solution Pipes

In plants where chlorine solution is piped, check on type of
pipe used. Fiberglass reinforced epoxy pipe has been
observed to erode inside through the action of the chlorine
solution. This poses a threat of contamination from exposed
glass fibers. Pipes made with polyester resin do not
deteriorate from this solution.

5.4.5.8 - Sanitation Practices

Observe sanitizing practices throughout the plant and
evaluate their effectiveness, degree of supervision
exercised, strength, time, and methods of use of sanitizing
agents. Determine the use, or absence of, sanitizing
solutions both for sanitizing equipment and utensils as well
as for hand dipping. If chlorine is used, 50 ppm - 200 ppm
should be used for equipment and utensils, while a 100
ppm will suffice for hand dipping solutions. Sanitizing

solutions rapidly lose strength with the addition of organic material. The strength of the solution should be checked several times during the inspection.

5.4.6 - Manufacturing Process

Where helpful to describe equipment and processes, draw flow plans or diagrams to show movement of materials through the plant. Generally a brief description of each step in the process is sufficient. List all quality control activities for each step in the process and identify Critical Control Points. Provide a full description when necessary to describe and document objectionable conditions, or where the assignment specifically requests it.

Observe whether hands and equipment are washed or sanitized after contact with unsanitized surfaces. For example:

1. Workers do general work, then handle the product;

2. Containers contact the floor, then are nested or otherwise contact product or table surfaces;

3. Workers use common or dirty cloths or clothing for wiping hands;

4. Product falls on a dirty floor or a floor subject to outside foot traffic and is returned to the production line.

Be alert for optimum moisture, time and temperature conditions conducive to bacterial growth.

In industries where scrap portions of the product are re-used or re-worked into the process (e.g., candy and macaroni products), observe the methods used in the re-working and evaluate from a bacteriological standpoint. Re-working procedures such as soaking of macaroni or noodle scrap to soften or hand kneading of scrap material offers an excellent seeding medium for bacteria.

When a product is processed in a manner which destroys micro-organisms, note whether there are any routes of recontamination from the "raw" to the processed product (e.g. dusts, common equipment, hands, flies, etc.).

5.4.6.1 - Ingredient Handling

Observe the method of adding ingredients to the process. Filth may be added into the process stream from dust, rodent excreta pellets, debris, etc. adhering to the surface of ingredient containers. Evaluate the effectiveness of cleaning and inspectional operations performed on the materials prior to or while adding to the process. Determine specific trimming or sorting operations on low quality or questionable material. Observe and report any significant lags during the process or between completion of final process and final shipping. For example, excessive delay between packing and freezing may be a factor in production of a violative product.

5.4.6.2 - Formulas

The Act does not specifically require management to furnish formula information except for human drugs, restricted devices and infant formulas. Nonetheless, they should be requested especially when necessary to document violations of standards, labeling, or color and food additives. Management may provide the qualitative formula but refuse the quantitative formula.

If formula information is refused, attempt to reconstruct formula by observing:

1. Product in production,

2. Batch cards or formula sheets,

3. Raw materials and their location.

Any refusal to furnish requested information is reported in your EIR under the refusal heading.

5.4.6.3 - Food Additives

Refer to the food additives programs in CPGM (Chapter 9) for instructions on conducting establishment inspections of firms manufacturing food additive chemicals. Information is also available in DFI's "Guide to the Inspection of Manufacturers of Miscellaneous Food Products - Volume 2.

When making food plant inspections direct your evaluation of food additives only to those instances of significant violation or gross misuse.

Routine inspectional coverage will be directed primarily to the following two types of additives:

1. Unauthorized and illegal as listed in the Food Additive Status List (safrole, thiourea, et al), and

2. Restricted as to amount in finished food.

Because of special problems, exclude the following additives from coverage during routine inspections:

1. Packaging materials,

2. Waxes and chemicals applied to fresh fruit and vegetables,

3. Synthetic flavors and flavoring components except those banned by regulations or policy statements (these products will be covered under other programs), and

4. Food additives in feeds (these products will be covered under other programs).

The Food Additives Status List (FASL) found on the CFSAN website contains an alphabetical listing of substances, which may be added directly to foods or feeds and their status under the Food Additives Amendment and Food Standards. In addition, a few unauthorized or illegal substances are included.

You may encounter substances not included in the Food Additives Status List (FASL). Such substances will include:

1. Obviously safe substances not on the list of items generally recognized as safe (GRAS), which are not published in the regulations, i.e., salt, cane sugar, corn syrup, vinegar, etc.;

2. Synthetic flavoring substances because of their indefinite status;

3. Substances pending administrative determination,

4. Substances granted prior sanction for specific use prior to enactment of the Food Additives Amendment.

Give primary attention to unauthorized substances. Document and calculate levels of restricted-use additives in finished food only where gross misuse or program violations are suspected as follows:

1. List ingredients, which may be restricted substances or food additives, and determine their status by referring to the current FASL. Report complete labeling on containers of these substances.

2. Obtain the quantitative formula for the finished product in question.

3. Determine the total batch weight by converting all ingredients to common units.

4. Calculate the theoretical levels in the final product of all restricted or unauthorized ingredients from the formula by using the Food Additives Nomographs. See IOM Exhibit 5-11.

5. Determine probable level of restricted ingredients by observing the weight of each ingredient actually put into the batch.

5.4.6.4 - Color Additives

Evaluate the status of color additives observed during each establishment inspection by using the Color Additive Status List and the Summary of Color Additives Listed in the United States in Food, Drugs, Cosmetics, and Medical Devices. Both of these links can be found on the CFSAN website. These lists provide the current status and use limitations of most color additives likely to be found in food, drug, device, or cosmetic establishments.

Stocks of delisted and uncertified colors may be found in the possession of manufacturers where there is no evidence of misuse. Advise the firm of the status of these colors. If management wishes to voluntarily destroy such colors, witness the destruction and include the facts in your EIR. If the firm declines to destroy the colors, determine what disposition is planned, i.e., use in non-food, non-drug, non-cosmetic or non-medical device products. The validity of certification information can be checked by accessing the online Color Certification Database system maintained by the Office of Cosmetics and Colors or contact Ray Decker,

Director, Division of Color Certification and Technology, HFS-105, by e-mail at raymond.decker@fda.hhs.gov to be granted user privilege.

Where decertified or restricted-use colors are used in manufacturing food, drug, device, or cosmetics products, proceed as follows:

1. Collect an Official Sample consisting of the color and the article in which it is being used. Make every effort to collect interstate shipments of the adulterated product before attempting to develop a 301(k) or 301(a) case. When regulatory action is an alternative, obtain sufficient interstate records to cover both the color and the basic ingredients of the manufactured product. Refer to IOM Sample Schedule, Chart 9 - Sampling Schedule for Color Containing Products for guidance.

2. Document the use of decertified colors after the decertifying date. Documentation should include batch formula cards, employee statements, code marks indicating date of manufacture, color certification number, etc. The presence of color in the finished product will be confirmed by your servicing laboratory.

5.4.6.5 - Quality Control

The objective of quality control is to ensure the maintenance of proper standards in manufactured goods, especially by periodic random inspection of the product. Your inspection should determine if the firm's quality control system accomplishes its intended purpose. Establish responsibility for specific operations in the control system. Determine which controls are critical for the safety of the finished product.

5.4.6.5.1 - Inspection System

Determine what inspectional control is exercised over both raw materials and the processing steps. Such inspection may vary from simple visual or other organoleptic examination to elaborate mechanical manipulation. Determine what inspection equipment is used, i.e., inspection belts, sorting belts, grading tables, ultraviolet lights, etc. Ascertain its effectiveness, maintenance or adjustment schedules. Where indicated, determine the name of the manufacturer of any mechanical inspection device and the principles of its operation.

Evaluate the effectiveness of the personnel assigned to inspection operations. Determine if the inspection belts or pick-out stations are adequately staffed and supervised.

Determine the disposition of waste materials, which are unfit for food or feed purposes.

5.4.6.5.2 - Laboratory Tests

Describe routine tests or examinations performed by the firm's laboratory and the records maintained by the firm. Determine what equipment is available in the laboratory and if it is adequate for the purpose intended. If the firm uses a consulting laboratory, determine what tests are performed and how often. Review laboratory records for the period immediately preceding the inspection.

5.4.6.5.3 - Manufacturing Code System

Obtain a complete description of the coding system with any necessary keys for interpretation. Provide an example by illustrating the code being used at the time of the inspection. (See 21 CFR 113.60(c) and 114.80(b)). Report coding systems, which require the use of ultra-violet light for visibility. Hermetically sealed containers of low acid processed food must be coded in a manner clearly visible. (See 21 CFR 113.60). Check 21 CFR 113 and 114 for regulations on coding for the type plant you are inspecting.

5.4.6.6 - Packaging and Labeling

Evaluate storage of packaging materials including protection from contamination by rodents, insects, toxic chemicals or other materials. Appraise the manner in which containers are handled and delivered to the filling areas. Determine if there is likelihood of chipping of glass or denting, puncturing, tearing, etc., of packaging materials. Observe the preparation of containers prior to filling. Consider any washing, steaming, or other cleaning process for effectiveness. Determine, in detail, the use of air pressure or other cleaning devices.

5.4.6.6.1 - Quantity of Contents

If slack fill is suspected, weigh a representative number of finished packages. See IOM 4.3.8 for net weight procedure. Sets of official weights are available in the district servicing laboratory. These may be used to check the accuracy of firm's weighing equipment.

5.4.6.6.2 - Labeling

Check the sanitary condition of labelers and equipment feeding cans to, and away from, the labeler. Determine if old product is present on any equipment which touches the can end seams, in the presence of moisture carry-over from the can cooling operation. Check availability of floor drains in the labeling area. Absence of floor drains could indicate infrequent cleaning of the equipment unless it is physically moved to another area for cleaning.

Determine what labels are used and what labeling is prepared or used to accompany or promote the product. Obtain specimens of representative labels and labeling including pamphlets, booklets, and other promotional material. Obtain 3 copies of labels and labeling believed to be violative.

5.4.6.6.3 – Nutritional and Allergen Labeling

See document "Guide to Nutritional Labeling and Education Act (NLEA) Requirements" for guidance. See http://www.cfsan.fda.gov/~dms/wh-alrgy.html for guidance on Food Allergen Labeling and Consumer Protection Act of 2004 (FALCPA) requirements.

5.4.7 - Sanitation

Documented observation of the conditions under which food products are processed, packed, or stored is essential to the proper evaluation of the firm's compliance with the law. This involves the determination of whether or not insanitary conditions contribute to the product being adulterated with filth, rendered injurious to health, or whether it consists in whole or in part of a filthy, putrid or decomposed substance.

Observations that dirt, decomposed materials, feces or other filthy materials are present in the facility and there is a reasonable possibility these filthy materials will be incorporated in the food are also ways of determining products may have become contaminated.

5.4.7.1 - Routes of Contamination

It is not sufficient to document only the existence of insanitary or filthy conditions. You must also demonstrate how these conditions contribute or may contribute to contaminating the finished product. Investigate and trace potential routes of contamination and observe all means by which filth or hazardous substance may be incorporated into the finished product. For example, defiled molding starch in a candy plant may contribute filth to candy passing through it, or filth in insect or rodent contaminated raw materials may carry over into the finished product. IOM Section 4.3.7 contains instructions on sample collection techniques for adulteration violations, including instructions for field exams and sample collections to document evidence of rodent, insect, etc., contaminated lots, and instructions for in-line sampling, including bacteriological samples. Finished product sample sizes for filth and micro collections can be found in the applicable Compliance Program (CPGM) or DFI "Guide to Inspections of ***."

5.4.7.1.1 - Insects

Insect contamination of the finished product may result from insect infested raw material, infested processing equipment or insanitary practices, and by insanitary handling of the finished product. When routes of contamination with insect filth are encountered, identify the insects generally, e.g., weevils, beetles, moths, etc. If qualified, identify as to species. You must be correct in your identification. See IOM Appendix A.

5.4.7.1.2 - Rodents

Rodent contamination of the finished product may result from using rodent defiled raw materials, exposure to rodents during processing, and by rodent depredation of the finished product. When evidence of rodents are discovered, you should thoroughly describe its composition, quantity, estimated age and location. Explain its significance and potential for product contamination.

5.4.7.1.3 - Pesticides

Pesticide contamination of the finished product may be the result of mishandling of food products at any stage in manufacturing or storage. The use of toxic rodenticides or insecticides in a manner, which may result in contamination, constitutes an insanitary condition. Where careless use of these toxic chemicals is observed, take photographs and provide other documentation showing its significance in relation the food products.

Additional guidance can be found in 21 CFR as follows:

1. Part 110.20(b) - Plant Construction and Design,

2. Part 110.40(a) - Equipment and Utensils,

3. Part 110.35(c) - Pest Control,

4. Part 110.10(b) - Personnel Cleanliness.

Additional guidance can be found in 40 CFR Part 180 - Tolerances and Exemptions From Tolerances For Pesticides in Food Administered by The Environmental Protection Agency as follows:

1. Part 180.521 - Fumigants for grain-mill machinery; tolerances for residues, and

2. Part 180.522 - Fumigants for processed grains used in production of fermented malt beverages; tolerances for residues.

Be alert for:

1. Possible PCB contamination. Articles containing PCBs (e.g., transformers, PCB containers stored for disposal, electrical capacitors) must be marked with prescribed labeling to show they contain PCBs. No PCB-containing heat exchange fluids, hydraulic fluids or lubricants are allowed used in food plants. All PCB storage areas must be marked to show the presence of PCBs. Observe food plant transformers for possible leakage. If observed, determine if food items are stored in the area, and sample for PCB contamination. If PCBs are encountered in a food establishment, immediately advise management this is an objectionable condition and advise your supervisor.

2. Possible mix-up of pesticides or industrial chemicals with food raw materials.

3. Improperly stored pesticides or industrial chemicals (lids open, torn bags in close proximity to foods, signs of spillage on floors, pallets, shelves, etc.).

4. Incorrect application methods including excessive use. Many pesticide labels give instructions for use and precautions on the container.

5. Improper disposal or reuse of pesticide or industrial chemical containers.

6. Evidence of tracking powder or improper use of bait stations or baited traps.

7. Improper handling of equipment. Movable or motorized equipment used for handling possible chemical contaminants should not be used for handling food products unless they are thoroughly decontaminated. For example, fork-lifts moving pallets of pesticides should not also be used to move pallets of flour, etc.

8. Use of unauthorized pesticides.

9. Use of foods treated with pesticides and marked "Not For Human Consumption" (e.g., Treated seed wheat, etc.).

10. Noticeable odor of pesticides.

11. Careless use of machinery lubricants and cleaning compounds.

12. Chemical contaminants in incoming water supply.

When inspecting products with a known potential for metals contamination, determine whether the firm tests for such contamination in raw materials.

Determine who administers the firm's rodent and insect control program. Determine responsibility for the careless use of toxic materials.

If pesticide misuse is suspected, obtain the following information;

1. Name of exterminator and contract status,

2. Name of pesticide,

3. Name of pesticide manufacturer,

4. EPA registration number,

5. Active ingredients, and

6. Any significant markings on pesticide containers.

Fully document the exact nature of any pesticide or industrial chemical contamination noted or suspected. If samples are collected to document misuse, exercise caution to prevent contamination of the immediate area of use, product or yourself.

5.4.7.1.4 - Other

Contamination of food products by bats, birds and/or other animals is possible in facilities where food and roosting facilities are available. Examine storage tanks, bins, and warehousing areas to determine condition and history of use. There have been instances where empty non-food use containers were used for food products.

5.4.7.2 - Microbiological Concerns

During the inspection, you must fully identify likely sources and possible routes of contamination of the product. See IOM 4.3.7.7 for instructions on sampling for pathogens. Before and during processing, become completely familiar with the flow of the process and determine the potential trouble spots, which may be built into the operation. To document the establishment is operating in an insanitary manner, it is necessary to show the manufacturer has contributed to the bacterial load of the product. If there are several products being prepared at once, do not try to cover the entire operation during one inspection. Select the product which has the greatest potential for bacterial abuse or which poses the greatest risk for the consumer.

It is extremely important each EIR contain complete, precise, and detailed descriptions of the entire operation. The EIR must be able to stand alone without the analytical results, which serve to support the observations.

Observations made during the inspection must be written in clear and concise language. The EIR will be reviewed in conjunction with analytical results of in-line, environmental and finished product samples. Based on this review and other information, which may be available, the district must then decide if the total package will support a recommendation for regulatory action.

Each inspection/process will be different, but the techniques for gathering the evidence will be the same. However, the critical points in the operation should always be defined and special attention given to these areas.

Depending on the type of product being produced and the process being used, it may be useful to record the time each critical step takes from beginning to end of the entire processing period with correlating temperature measurements. This should be done especially for products, which would support the growth of microbial pathogens. During the entire inspection, be aware of and document delays in the processing of the product (e.g., temperature of product prior to, during and after the particular processing step, and the length of time the product has been delayed prior to the next step).

Some products receive a thermal process at the end of production, which may reduce bacterial counts to or near zero. Include detailed observations of heating step, temperature, length of time, controls and documentation used/not used by the firm. Even in the presence of end-product thermal processing, there is a regulatory significance to insanitary conditions prior to cooking, coupled with increases in bacterial levels demonstrated through in-line sampling.

5.4.7.2.1 - Processing Equipment

Document the addition, or possible addition of pathogenic microorganisms from accumulated material due to poorly cleaned and/or sanitized processing equipment

Observe and report the firm's clean up procedures and the condition and cleanliness of food contact surfaces before production starts, between production runs and at the end of the day. Document any residue on food contact surfaces of equipment, especially inside complex equipment not easily cleaned and sanitized. Report firms clean-up procedures in depth, since it may lend significance to insanitary conditions of residues on the plant machinery which are left to decompose overnight or between shifts. Where possible, observe equipment both before and after cleaning to assess it adequacy. Observations of residues on plant machinery can dramatically document the addition of pathogenic microorganisms, if present, into the product.

Identify any vectors of contamination (e.g. birds, rodents, insects, foot traffic, etc.), and describe sources and the routes of contamination from them to the product. Support this with your actual observations.

5.4.7.2.2 - Employee practices

Document any poor employee practice and how they have or would provide a route for contaminating the product. For example, did employees (number/time of day) fail to wash and sanitize their hands at the beginning of processing, after breaks, meals, or after handling materials likely contaminated with a microbial pathogen, etc.; and then handle the finished product. Did employees handle product in an insanitary manner (cross contaminating raw product with cooked product, etc., how many, how often).

5.4.7.3 - Storage

Evaluate the storage of finished products in the same manner as for raw materials. Determine if products are stored to minimize container abuse, facilitate proper rotation, and adherence to the storage requirements. This includes refrigeration temperatures, critical temperature tolerance, aging of products, and proper disposition of distressed stock.

5.4.7.3.1 - Food Transport Vehicles

During food sanitation inspections, (See IOM 5.2.2.2 regarding issuance of FDA 482, Notice of Inspection while inspecting vehicles.), conduct inspections of food transport vehicles to include:

1. Evidence of insanitary conditions,

2. Conditions which might lead to food adulteration,

3. Physical defects in the vehicle,

4. Poor industry handling practices.

The following types of transport vehicles should be covered:

1. Railroad boxcars, both refrigerated and non-refrigerated, and hopper cars.

2. Any type of truck used to transport foods; both refrigerated and non-refrigerated.

3. Use extreme caution, if it is necessary to inspect tank railcars or tank trucks. Usually this coverage will be limited to determining what was transported in the tank previously and was the tank cleaned and/or sanitized as necessary between loads.

4. Vessels used to transport food in I/S commerce. Direct coverage primarily to intercoastal type vessels, including barges.

Coverage should be limited to food transport vehicles used for long haul (I/S) operations. Long haul vehicles are defined as those which travel at least 150 miles between loading and unloading or which do not return to the point of loading at the end of the day.

Regulatory actions are possible if unfit cars are loaded and, as a result of loading, adulteration occurs. Fully document any violations noted with appropriate samples and photographs. When vehicle insanitation is observed, it is imperative the carrier's and shipper's responsibility for the food adulteration be documented by appropriate evidence development, such as:

1. The nature and extent of the conditions or practices, and

2. The mechanical or construction defects associated with the food transport vehicle.

3. Individual responsibility for vehicle or trailer cleaning, vehicle assignments, load assignments, etc.

If gathering evidence about a single carrier, seek a series of occurrences at numerous locations involving as many different shippers as possible.

Basically two types of vehicles will be covered.

5.4.7.3.2 - Vehicles at Receivers

When inspecting receivers of food products, examine the food transport vehicle prior to or during unloading. Make a preliminary assessment of food product condition, then inspect the vehicle after unloading to determine its condition and whether the unloaded food may have been contaminated during shipment. If the food appears to have been adulterated, collect a sample(s) for regulatory consideration. Samples collected from vehicles, which have moved the product in interstate commerce are official samples. You may also collect Documentary (DOC) Samples from the vehicle to substantiate the route of contamination.

5.4.7.3.3 - Vehicles at Shippers

When inspecting shippers of food products, examine the food transport vehicle just prior to loading to determine its sanitary/structural conditions. If the vehicle has significant sanitation or structural deficiencies, notify the shipper of these conditions and of the possibility of product adulteration. If the shipper loads food aboard the vehicle, alert your supervisor so he/she can contact the FDA District where the consignee is located for possible follow-up. You may also collect samples from the load. These samples will become official when the Bill of Lading is issued.

5.4.8 - Distribution

Report the general distribution pattern of the firm. Review interstate shipping records or invoices to report shipment of specific lots. If access to invoices or shipping records is not possible, observe shipping cartons, loading areas, order rooms, address stencils, railroad cars on sidings, etc., to determine customer names, addresses and destination of shipments. If no products are suspect, obtain a listing of the firm's larger consignees.

5.4.8.1 - Promotion and Advertising

Determine the methods used to promote products and how the products reach the ultimate consumer. Determine what printed promotional materials are used and whether they accompany the products or are distributed under a separate promotional scheme. Check on the possibility of oral representations, i.e., door-to-door salesmen, spieler, etc. and obtain copies of brochures, pamphlets, tearsheets, instructions to salespersons, etc. Where indicated, obtain the lecture schedule of any promotional lecture program. If applicable, determine the general pattern of the media used for promotion and advertising.

5.4.8.2 - Recall Procedure

Determine the firm's recall procedure. Audit enough records to determine the effectiveness of established procedures. Report if there is no recall procedure.

5.4.8.3 - Complaint Files

Review the firm's complaint files. Where possible, copy the names and addresses of representative complainants; include a brief summary of each significant complaint in the EIR.

Identify who reviews complaints and their qualifications. Describe the criteria used by the firm in evaluating the significance of complaints and how they are investigated. Determine if records are kept of oral and telephone complaints. See IOM 5.2.8 for discussion of complaints with management and IOM 5.10.4.3.11 for reporting of complaints in the EIR.

Complaints may not be filed in one specific file, but may be scattered throughout various files under other subject titles including Product name; Customer name; Injured party name; Adjustment File; Customer Relations; Repair orders, etc.

During the inspection investigate all complaints contained on FDA-2516 and FDA-2516a forms in the firm's district factory jacket. See IOM 5.2.8, 5.4.1.1 and 5.10.4.3.11.

5.4.9 - Other Government Inspection

See IOM 3.1 for general procedures on cooperating with other Federal, State, and local officials.

During Establishment Inspections determine the specific type of inspection service and inspecting units, which cover the firm, such as the name of the federal, state, county, or city health agency or department. Obtain the name and title of the inspectional official, and general method of operation.

5.4.9.1 - Federal

Do not inspect firms, or those portions of the plant, subject to compulsory, continuous inspection under USDA's Meat Inspection Act, Poultry Products Inspection Act, or Egg Products Inspection Act, except on specific instructions from your supervisor or assignment document.

Ingredients or manufacturing processes common to both USDA and FDA regulated products should be inspected by FDA. See IOM 3.2.1.3 for FDA-USDA Agreements in specific areas.

Provide routine FDA coverage of such firms as breweries and wineries, which may be intermittently inspected on a compulsory basis by the U.S. Treasury Department, U.S. Public Health Service, or other agencies.

All products inspected under the voluntary inspection service of the Agriculture Marketing Service (AMS), USDA, and the National Marine Fisheries Service (NMFS), US Department of Commerce, are subject to FDA jurisdiction and are usually given routine coverage. However, formal written Agreements or Memoranda of Understanding between FDA and other agencies are often executed and may govern the agreeing agencies' operations on this type of inspected plants. When assigned this type of plant for inspection, always check to see if an Agreement or a Memorandum of Understanding exists between FDA and the agency involved to determine the obligations of both agencies. See IOM 3.1.2.1 and 3.2.

If you are assigned to cover a Federally Inspected plant which is under either compulsory or voluntary inspection, present your credentials and an FDA 482 "Notice of Inspection" to management and:

1. Identify yourself to the inspector(s) and invite him/her to accompany you on the inspection but do not insist on their participation.

2. At the conclusion of the inspection, offer to discuss your observations and provide the in-plant inspector with a copy of your Inspectional Observations (FDA 483).

5.4.9.2 - State and Local

State and local officials usually have extensive regulatory authority over firms in their area regardless of the interstate

movement or origin of the food products involved. Joint
FDA-State or local inspections are frequently conducted.
These are usually arranged by district administrative or
supervisory personnel. See IOM 3.1.2 and 3.3.

5.4.9.3 - Grade A Dairy Plant Inspections

If you are assigned to do an inspection or sample collection
at a dairy firm in the Grade A program or a firm which has
products labeled and sold as Grade A, you should verify the
need to complete the assignment with your supervisor and
the Regional Milk Specialist. Grade A plants and most
products labeled as Grade A are inspected by state
inspectors or FDA's Regional Milk Specialists and you
should not inspect these products. Firms in the Grade A
program and covered by the Interstate Milk Shippers (IMS)
program are identified in the Interstate Milk Shippers List of
Sanitation Compliance and Enforcement Ratings book
published every year by the Milk Safety Branch, HFS-626.
The reference lists the specific plant and each product
covered under the IMS program. These products are
covered by a MOU between the FDA and the states, which
places primary inspectional responsibility with the state.

There are situations where you will need to conduct an
inspection in a Grade A plant and cover products they
manufacture which do not carry the "Grade A" designation
(such as juices). If the plant is an IMS shipper and has fluid
or other products rated as acceptable they may also
manufacture optional products and label them as Grade A,
without having those product lines covered in the IMS
program. Fluid products, sour cream, cultured milk products
and yogurt are not optional products. Optional products
include cottage cheese, condensed milk/whey and dry
milk/whey and may be labeled as Grade A. As an example,
some firms are listed in the IMS Sanitation Compliance and
Enforcement Ratings book who manufacture cottage
cheese and label it as Grade A, but is not specifically
covered under the Grade A inspection program. As long as
the plant is listed and has one or more products rated as
acceptable in the Grade A program, they can manufacture
a product and label it as Grade A without having the
particular product line covered in the IMS program. In those
situations, the product will not be shown in the Enforcement
Ratings book and you should cover its production, labeling,
etc.

5.4.10 - Food Standards

The Federal Food, Drug, and Cosmetic Act requires the Secretary of Health and Human Services to promulgate reasonable definitions and Standards for food to promote honesty and fair dealing in the interest of consumers. When a Standard becomes effective, it establishes the common or usual name for the article, defines the article and fixes its standard of identity. It is then the official specification for the food. The food industry actively participates in the development of a Standard, and supplies much of the data upon which the regulation is based.

The Food Standards (FS) Inspection is made to obtain data for use, together with information from other sources in developing a Food Standard. Food Standard inspections are also made to determine a firm's compliance with food standards regulations, when manufacturing a standardized food.

5.4.10.1 - Food Establishment Inspection

Food Standard (FS) inspection assignments usually originate from CFSAN. When an inspection is planned for the purpose of collecting data to support a proposed food standard regulation, the district may elect to advise the firm, if the CFSAN has not already done so. If the firm selected does not choose to cooperate, it may be necessary to visit additional plants in order to obtain the desired information. Selection of additional firms should be done in consultation with the CFSAN.

Some firms often contend their entire process and formulas are "trade secrets". Attempt to persuade management the term "trade secret" should only be used to cover the process and/or quantitative-qualitative formulation which is truly unique to the firm. In instances where the firm is reluctant to release any of the information requested, point out FDA will, within the limits of the Freedom of Information Act, make every effort to preserve the confidentiality of the composition, make-up, and production levels of the product through the use of codes, which cannot be traced back to the firm. Include as much of the compositional and processing information as you can in the body of the report, without violating the firm's confidence.

5.4.10.2 - Food Inspection Report

FS EIR's may be used as exhibits at public hearings and are subject to review by any interested party.

Three copies of the report are prepared. The original and one copy will be submitted to the CFSAN and one copy kept for the district file. Sign the original and duplicates of the first and last pages of each report sent to the Center.

Divide the report into three sections.

5.4.10.2.1 - Establishment Inspection Record (EI Record)

In order to relate the sections of the report to each other and to any assignments, and to assure any parts of the reports made public will not be identified as to the name of the firm or individuals therein, each district will set up a master list of numbers. One number will be assigned to each establishment covered, e.g., "BLT FS-3". For each FS Inspection place the assigned number next to the firm name on the EI Record. All other pages of the report shall be identified only by this number, the name of the commodity, and date. Example: "EIR Frozen Fish Sticks 10-3-87 BLT FS-3". This indicates a FS EI of frozen fish sticks conducted by Baltimore District on 10-3-87 in a plant designated as #3.

Where a producer may be reluctant to release any of the information requested, point out the FDA will, within the limits of the FOIA, make every effort to preserve the confidentiality of the composition, make-up, and production levels of his product through the use of codes, which cannot be traced back to the firm.

5.4.10.2.2 - Body of Report

Prepare the body of the report following the narrative outline as for any other food EIR except for the restrictions below.

The body of the FS report should also contain information in regard to the approximate annual value and volume as well as the percent of interstate business for each product covered. This is necessary because the coversheet, which contains this information, identifies the firm and will not be made public. Processes and the listing of raw materials used by the firm, which are not restricted by the term "trade secret" should be included. Any opinions, recommendations, or other information obtained or offered by individuals interviewed should be reported. Any suggestions made by individuals interviewed regarding what should be placed in the Standards for the products covered should be included. All individuals interviewed, firm name, etc. should have an identifying code assigned.

The body of the report should not include names and titles of individuals, (including USDA, USDI, or other inspectors), trade secret information, labeling, trade names, formulas, sample numbers, firm name or location of plant (other than by state or region), shipments, or other distribution information, legal status, or regulatory history. This information will be placed in the "Special Information" section of the report.

5.4.10.2.3 - Special Information Section

This is a separate attachment to the EIR which lists the names and titles of individuals (including other government inspectors) and firms with a reference code for each. The EIR should refer only to "Mr. A.," "Mr. B.," "Firm X," "Firm Y", etc. Do not use the firm or individual's actual initials in the body of the report. Include all information excluded from the body of the report and mount all labels obtained during the EI Labels may be quoted in the body of the report, but do not identify the firm. List the "Special Information Sheet" in the FACTS endorsement section as an enclosure.

Supplemental Reports - If, because of an additional visit or visits to the same firm on the same project, it is necessary to prepare another EIR, flag the report with the same number as assigned to the original report. For example, mark the EI Record "BLT FS-3 Supplemental Report", and the remaining pages, "EIR Frozen Fish Sticks 10-25-87 BLT FS-3 Supplemental Report."

5.4.10.3 - Violative Inspections

When an inspection made in connection with the Food Standards project shows insanitary or other conditions which are not germane to the assignment or in the District's opinion suggests regulatory action, an appropriate narrative of the violative conditions should be prepared as a Regulatory Addendum.

5.5 - Drugs

5.5.1 - Drug Inspections

Authority for inspection is discussed in IOM 2.2. FD&C Act
Sections 501(a)-(d) [21 U.S.C. 351(a)-(d)] describe the
ways in which a drug may be or may become adulterated.
Section 502 of the FD&C Act [21 U.S.C. 352] does the
same, with respect to misbranding. Section 505 of the
FD&C Act [21U.S.C. 355] requires that new drugs be
approved by FDA. Therefore, the purposes of a drug
inspection are:

1. To determine whether a firm is distributing drugs that
   lack required FDA approval including counterfeit or
   diverted drugs;

2. To determine and evaluate a firm's adherence to the
   concepts of sanitation and good manufacturing
   practice;

3. To assure production and control procedures include all
   reasonable precautions to ensure the identity, strength,
   quality, and purity of the finished products;

4. To identify deficiencies that could lead to the
   manufacturing and distribution of products in violation of
   the Act, e.g., non-conformance with Official Compendia,
   super/sub potency, substitution;

5. To obtain correction of those deficiencies;

6. To determine if new drugs are manufactured by the
   same procedures and formulations as specified in the
   New Drug Application documents;

7. To determine the drug labeling and promotional
   practices of the firm;

8. To assure the firm is reporting NDA field alerts as
   required by 21 CFR 314.81 and Biological Product
   Deviation Reports (BPDRs) for therapeutic biological
   products as required by 21 CFR 600.14;

9. To determine if the firm is complying with the requirements of the Prescription Drug Marketing Act (PDMA) and regulations; and

10. To determine the disposition of Drug Quality Reports (DQRS) received from the Division of Compliance Risk Management and Surveillance/CDER; and

11. To determine if the firm is complying with postmarket Adverse Drug Experience reporting requirements as required by 21 CFR sections 310.305 (prescription drugs without approved NDA/ANDA), 314.80, 314.98, and 314.540 (application drug products), and 600.80 (therapeutic biological products), and Section 760 of the FD&C Act (non-application nonprescription products) [21 U.S.C. 379aa].

5.5.1.1 - Preparation and References

Become familiar with current programs related to drugs. Determine the nature of the assignment, i.e., a specific drug problem or a routine inspection, and if necessary, consult other district personnel, such as chemists, microbiologists, etc., or center personnel, such as office of compliance staff. Review the district files of the firm to be inspected including:

1. Establishment Inspection Reports,

2. District Profiles,

3. Drug Applications (New, Abbreviated and Investigational),

4. Therapeutic Biologics License Applications,

5. Sample results,

6. Complaints and recalls,

7. Regulatory files,

8. Drug Quality Reports (DQRs), NDA Field Alert Reports (FARs), and Biological Product Deviation Reports (BPDRs),

9. Drug Registration and Listing

During this review identify products which:

1. Are difficult to manufacture,

2. Require special tests or assays, or can not be assayed,

3. Require special processes or equipment, and

4. Are new drugs and/or potent low dosage drugs.

Review the factory jacket, FACTS OEI and registration/listing data, and all complaint reports which are marked follow-up next inspection. These complaints are to be investigated during the inspection and discussed with management. See IOM 5.2.7.

Become familiar with current regulations and programs relating to drugs, CPGM 7356.002, et al. When making GMP inspections, discuss with your supervisor the advisability of using a microbiologist, analyst, engineer, or other technical personnel to aid in evaluating those areas of the firm germane to their expertise. Review the FD&C Act, Chapter V, Drugs and Devices. Review parts of 21 CFR 210/211 applicable to the inspection involved and Bioavailability (21 CFR 320). In the case of APIs, review FD&C Act section 501(a)(2)(B) [21 U.S.C 351(a)(2)(b)] and the ICH industry guideline entitled "Q7A Good Manufacturing Practice Guidance for Active Pharmaceutical Ingredients."

Review the current editions of the United States Pharmacopeia (USP), and Remington's Pharmaceutical Sciences for information on specific products or dosage forms. Also IOM 1.10.3 provides a link to a consolidated list of pertinent guides and guidelines which may be applicable during drug inspections.

Review 21 CFR 203 "Prescription Drug Marketing", 21 CFR 205 "Guidelines for State Licensing of Rx Drug Distributors", and CPGM 7356.022, Enforcement of the Prescription Drug Marketing Act (PDMA).

Before conducting an inspection that may involve postmarketing adverse drug experience reporting, you should review 21 CFR Sections 310.305, 314.80, 314.98, 314.540, and 600.80, Section 760 of the FD&C Act [21 U.S.C. 379aa], CPGM 7353.001, and the training video, 'Field Investigators: Adverse Drug Effects (ADE)

Detectives,' available online at http://www.fda.gov/cder/learn/ADE/ADE_Pagex.htm.

The Division of Manufacturing and Product Quality (DMPQ) in CDER has established two mechanisms for you to obtain technical assistance before, during, or after an inspection:

1. Division of Manufacturing and Product Quality (DMPQ) Subject Contacts (http://www.fda.gov/cder/dmpq/subjcontacts.htm#subjects). This list contains the names and phone numbers of DMPQ individuals identified as technical specialists in various areas.

2. Questions and Answers on Current Good Manufacturing Practices for Drugs (http://www.fda.gov/cder/20guidance/cGMPs/default.htm). This forum is intended to provide timely answers to questions about the meaning and application of CGMPs for human, animal, and biological drugs, and to share these widely. These questions and answers generally clarify statements of existing requirements or policy.

5.5.1.2 - Inspectional Approach

Follow Compliance Program Guidance Manual (CPGM) 7356.002 and others as appropriate when conducting CGMP inspections. In-depth inspection of all manufacturing and control operations is usually not feasible or practical. A risk-based systems audit approach is recommended in which higher risk, therapeutically significant, medically necessary, and difficult to manufacture drugs are covered in greater detail during an inspection. The latter group includes, but is not limited to, time release and low dose products, metered dose aerosols, aseptically processed drugs, and formulations with components that are not freely soluble.

CPGM 7356.002 incorporates the systems-based approach to conducting an inspection and identifies six (6) systems in a drug establishment for inspection: Quality, Facilities and Equipment, Materials, Production, Packaging and Labeling and Laboratory Control Systems. The full inspection option includes coverage of at least four (4) of the systems; the abbreviated inspection option covers of at least two (2) systems. In both cases, CPGM 7356.002, indicates the Quality System be selected as one of the systems being covered. During the evaluation of the Quality System it is important to determine if top management makes science-based decisions and acts promptly to identify, investigate, correct, and prevent manufacturing problems likely to, or have led to, product quality problems.

When inspecting drug manufacturers marketing a number of drugs meeting the risk criteria, the following may help you identify suspect products:

1. Reviewing the firm's complaint files early in the inspection to determine relative numbers of complaints per product.

2. Inspecting the quarantine, returned, reprocessed, and/or rejected product storage areas to identify rejected product.

3. Identifying those products which have process control problems and batch rejections via review of processing trends and examining annual reviews performed under 21 CFR 211.180(e).

4. Reviewing summaries of laboratory data (e.g., laboratory workbooks), OOS investigations, and laboratory deviation reports.

5.5.1.3 - CDER Bio-research Monitoring

Bio-research monitoring (BIMO) assignments for drugs will generally be issued by the Center for Drug Evaluation and Research (CDER) (see IOM 5.5.6).

5.5.2 - Drug Registration & Listing

Registration and listing is required whether or not interstate commerce is involved. See Exhibit 5-12 and IOM 2.9.1.1 for additional information.

Two or more companies occupying the same premises and having interlocking management are considered one establishment and usually will be assigned a single registration number. See IOM 5.1.1.11 - Multiple Occupancy Inspections for additional information.

Independent laboratories providing analytical or other laboratory control services on commercially marketed drugs must register.

FACTS will indicate if the establishment is registered for the current year. If you determine registration and listing is required, advise your supervisor. After checking for past registration, cancellation, etc., the district will provide the firm with the proper forms and instructions.

Each establishment is required to list with FDA every drug in commercial distribution, whether or not the output of such establishment or any particular drug so listed enters interstate commerce. During the establishment inspection, you should remind the firm of its responsibilities for ensuring its drug listing accurately reflects the current product line and updating its listing as necessary to include all product changes, NDC changes, and discontinuations in accordance with 21 CFR 207. If registration and listing deficiencies are found, document it in your EIR, collect a documentary sample and/or contact your supervisor.

5.5.3 - Promotion and Advertising

21 CFR 202.1 which pertains only to prescription drugs, covers advertisements in published journals, magazines, other periodicals, and newspapers, and advertisements broadcast through media such as radio, television, and telephone communication systems. Determine what department or individual is responsible for promotion and advertising and how this responsibility is demonstrated. Ascertain what media (radio, television, newspapers, trade journals, etc.) are utilized to promote products.

Do not routinely collect examples of current advertising. Advertising should be collected only on assignment, or if, in your opinion, it is clearly in violation of Section 502(n) of the FD&C Act [21 U.S.C. 352 (n)] or 21 CFR 202.1.

5.5.4 - Guarantees and Labeling Agreements

Determine the firm's policies relative to receiving guarantees for raw materials, and issuing guarantees on their products. Also determine firm's practices regarding shipment of unlabeled drugs under labeling agreements. See IOM 5.3.7.2.

5.5.5 - Other Inspectional Issues

5.5.5.1 - Intended Use

Please see the discussion of jurisdiction in section IOM 5.10.4.3.6.

5.5.5.2 - Drug Approval Status

The investigator should ascertain whether the drugs manufactured by the firm are covered by an NDA, ANDA, OTC monograph, or marketed under a claim of DESI or "grandfather" status.

5.5.5.3 - Drug Status Questions

If you have questions about misbranding, new drug status, drug/cosmetic, or drug/food (dietary supplement) status, call the Division of New Drugs and Labeling Compliance in the CDER Office of Compliance (301-796-3110).

In rare cases, a drug may be unapproved and inappropriate for marketing under any circumstances (i.e., it cannot be reconditioned or reformulated into a product appropriate for marketing). If you encounter products in this category, contact your supervisor to determine if a CGMP inspection is warranted.

5.5.5.4 - Drug/Dietary Supplement Status

In instances where the drug/dietary supplement status of a product is unclear, the investigator should collect all related labeling and promotional materials including pertinent Internet web sites. This labeling and promotional material is often useful in determining the intended use of a product (See 21 CFR 201.128). Labeling, promotional materials and Internet web sites often contain information, for example, disease claims, that can be used to determine the intended use of a product and thereby if it is a dietary supplement or a drug and an unapproved new drug.

5.5.5.5 - Approved Drugs

Check the current programs in your CPGM, Section 505 of the FD&C Act [21 U.S.C. 355] and 21 CFR part 314 for required information. You may take the District's copy of the NDA into the plant as a reference during the inspection. Document and report all deviations from representations in the NDA even though they may appear to be minor.

5.5.5.6 - Investigational Drugs

Follow the instructions in pertinent programs in your CPGM or as indicated in the specific assignment received.

5.5.5.7 - Clinical Investigators and/or Clinical Pharmacologists

Inspections in this area will be on specific assignment previously cleared by the Administration. Follow guidance in the CPGM or assignment.

5.5.6 - CDER Bio-Research Monitoring

Inspectional activities in the bio-research monitoring (BIMO) programs involve all product areas and Centers, including In Vivo Bio-equivalence, Good Laboratory Practice (GLP) for Non-Clinical Laboratories, Institutional Review Boards (IRB), Sponsors, Monitors, Contract Research Organizations, and Clinical Investigators (CI). In most instances, inspections conducted under this program will be done on assignment from the respective Center and occasionally with the participation of Center personnel as part of the inspection team.

During team inspections with Center personnel, the Field Investigator is the team leader. See IOM 5.1.2.5. The Compliance Program Guidance Manual (CPGM) for each program provides a description of the program and detailed instruction for conducting inspections.

Districts will make the initial classification of inspections and the Center issuing the assignment will make the final decision after review.

5.5.7 - Adverse Event Reporting

FD&C Act section 760 [21 U.S.C. 379aa] and 21 CFR sections 310.305, 314.80, 314.98, and 314.540 require reporting of adverse events associated with the use of human drug products and section 600.80 requires reporting of adverse events associated with the use of biological products (including therapeutic biological products). Responsible firms include holders of NDAs and ANDAs, and manufacturers, packers and distributors that are named on the labels of all FDA approved drug products and all prescription drug products. Firms are required to develop written procedures and to maintain records related to adverse events, both foreign and domestic. Firms must evaluate adverse event data to determine if the event has had a serious outcome such as death, disability, hospitalization, or was a life threatening event, and if the event was expected (labeled) or unexpected (unlabeled) for the product. Responsible firms must submit adverse event information to FDA in expedited or periodic reports, as described in the regulations.

Refer to the Compliance Program Guidance Manual (CPGM) (section 7353.001) for the description of the program and for detailed instructions for conducting inspections.

5.5.8 - Drug Inspection Report

See IOM 1.1 English language requirement. The requirements in IOM 5.10.4.3, and any applicable Compliance Program Guidance Manuals can be used to help you prepare your report.

This does not cover the reporting requirements for a directed inspection with a narrow focus, such as a complaint follow-up or investigation into a recall. In those cases, use your judgment and guidance in IOM 5.10.4 about the depth of reporting required. Follow the instructions and format for a human drug inspection report as contained in IOM 5.10.4.2 and 5.10.4.3.

This human drug inspection report does not require full and detailed narratives for every area for every inspection. The firm's state of compliance, the previous inspectional report and information, complexity of operations and other aspects all are determinants in how much reporting will be necessary. In many cases, brief summaries addressing the format areas will be sufficient.

5.6 - Devices

5.6.1 - Device Inspections

See IOM 2.2 for discussion of statutory authority.

The term "device" is defined in Sec. 201(h) of the FD&C Act [21 U.S.C. 321 (h)]. In-vitro diagnostics (21 CFR 809) are devices, as defined in 201(h) of the Act [21 U.S.C. 321 (h)], and may also be biological products subject to Section 351 of the PHS Act.

Inspections involving devices should be made only by those individuals qualified by training and experience in the device area. Electronic product radiation is defined in 21 CFR 1000. Because of the specific nature of inspections and investigations involving radiation, only personnel who have special training in this field should be assigned such work. However, others may participate for training purposes. Specific Compliance Program Guidance Manuals designate the type of individual and special training required for work in these areas.

CAUTION: Radiation-emitting devices and substances present a unique hazard and risk potential. Every effort should be taken to prevent any undue exposure or contamination. Monitoring devices must be used whenever radiation exposure is possible. Investigators should also be on the alert for, and avoid contact with, manufacturing materials and hazards associated with the manufacturing of many types of devices, which may present a threat to health, e.g., ethylene oxide, high voltage, pathogenic biomaterials, etc. See IOM 1.5 for additional safety information.

5.6.1.1 - Technical Assistance

Each region and some districts have engineers and radiological health personnel available for technical assistance and consultation. Do not hesitate to make use of their services.

Engineers, quality assurance specialists, and expert investigators in ORA/ORO/Division of Field Investigations (DFI), HFC-130, 301-827-5653, are available for on-site consultation and assistance in problem areas. The division's subject matter experts are also available by telephone for consultation and to answer questions regarding regulation and program interpretation and QS/GMP application. Additionally, the CDRH Office of Compliance enforcement divisions (organized by device product) can be contacted as necessary.

WEAC has various personnel (biomedical, sterility, electronic, materials, mechanical, nuclear and plastics engineers) available for telephone consultation and on-site assistance. They can be reached at 617-729-5700.

5.6.1.2 - Sample Collection During Inspection

Because of the limited funds available for samples and the relatively high cost of device samples, it is essential you consider, in consultation with your supervisor, the following factors before collecting a physical sample of a device:

1. If follow-up to a QS/GMP deviation, will sampling demonstrate the deviation and/or a defective product? Documentary Samples may be more suitable for QS/GMP purposes.

2. Likelihood of the analysis showing the device is unfit for its intended use.

3. Samples costing over $250.00.

4. Laboratory capability to analyze the sample. See IOM 4.5.5.3.6 for sample routing information.

If you are still uncertain, discuss with your supervisor and contact the CDRH Laboratory or WEAC 781-729-5700 for assistance.

Contact CDRH for assistance as follows:

1. In-vitro Diagnostic Devices - Office of Science and Technology (HFZ-113).

NOTE: *Device samples do not require 702(b) portions. Include in the FDA 525 and with the C/R, if destined for different locations, a copy of the firm's finished device specifications, test methods and acceptance and/or rejection criteria.*

5.6.1.3 - Types of Inspections

General device inspections will be conducted under various Compliance Programs found in the Compliance Program Guidance Manual. The majority of these will be QS/GMP inspections, but often the reason for the inspection will vary. For example, inspections may be conducted to assist the pre-market clearance process (PMA or Class III 510(k)), to specifically address MDR concerns, or to assure in-depth coverage of an aspect of manufacturing (sterility). The following describes some of these inspections.

5.6.1.4 - CDRH Bio-research Monitoring

Bio-research monitoring (BIMO) assignments for medical devices will generally be issued by the Center for Devices and Radiological Health (CDRH) (see IOM 5.5.6).

5.6.2 - Medical Device Quality System/Good Manufacturing Practices

Section 520(f) of the FD&C Act [21 U.S.C. 360j(f)] provides the Agency with authority to prescribe regulations requiring that the methods used in, and the facilities and controls used for, the manufacture, packing, storage, and installation of medical devices conform to good manufacturing practices. The medical device Quality System/Good Manufacturing Practices Regulation (QS/GMP)(21 CFR 820) became effective on June 1, 1997.

21 CFR 820 is established and promulgated under the authority of Sections 501, 502, 510, 513, 514, 515, 518, 519, 520, 522, 701, 704, 801 and 803 of the FD&C Act (21 U.S.C. 351, 352, 360, 360c, 360d, 360e, 360h, 360i, 360j, 360l, 371, 374, 381 and 383). Failure to comply with the provisions of 21 CFR 820 renders a device adulterated under Section 501(h) of the FD&C Act [21 U.S.C. 351(h)].

The regulations promulgated under 21 CFR 820 establish minimum requirements applicable to finished devices, as defined in 820.1(a). This regulation is not intended to apply to manufacturers of components or parts of finished devices, but instead recommended to them as a guide. In some special cases, components have been classified as finished devices (dental resins, alloys, etc.) and are subject to the QS/GMP. Manufacturers of human blood and blood components are not subject to this part, but are subject to 21 CFR 606.

The QS/GMP includes regulations regarding Purchasing Controls, 21 CFR 820.50, Receiving, In-process and Finished Device Acceptance, 21 CFR 820.80, and Traceability, 21 CFR 820.65, that require finished device manufacturers exercise more control over the components they use in their devices. The preamble of the QS/GMP states: "Since FDA is not regulating component suppliers, FDA believes that the explicit addition to the CGMP requirements of the purchasing controls...is necessary to provide the additional assurance that only acceptable components are used." And "...inspections and tests, and other verification tools, are also an important part of ensuring that components and finished devices conform to approved specifications." It further states, "...traceability of components must be maintained so potential and actual problem components can be traced back to the supplier."

The medical device QS/GMP is an umbrella GMP that specifies general objectives rather than methods. It is left to the manufacturer to develop the best methods to meet these objectives. You must use good judgment in determining compliance with the QS/GMP, keeping in mind that it is an umbrella GMP and all requirements may not apply or be necessary. The purpose of the QS/GMP is to assure conformance to specifications and to ensure that all requirements that will contribute to assuring the finished device meets specifications are implemented. You should not insist that a manufacturer meet non-applicable requirements. Refer to IOM Exhibit 5-13 for types of establishments that are required to comply with the QS/GMP.

5.6.2.1 - Pre-Inspectional Activities

Prior to the start of any medical device inspection, the factory jacket or establishment history of the establishment should be reviewed. You should review the previous inspectional findings and subsequent correspondence between the establishment and FDA; any MDR or consumer complaints where it was determined follow-up would occur at the next inspection; and any notifications of recalls since the last inspection.

The following on-line databases should be queried through the CDRH Information Retrieval System (CIRS):

1. For Medical Device Reporting (MDR) data (MAUDE)

2. Registration and Listing data, and 510(k)

3. PMA summary data (OSCAR);

These databases are accessible to users with individual accounts. Accounts can be requested through the district or regional CIRS liaisons to DFI/Alan Gion 301-827-5649.

MDR data most useful in preparing for an inspection includes specific MDRs for the manufacturer (i.e., query by establishment's short name) for the time frame since the last inspection, or MDRs for the generic devices manufactured by that establishment (i.e., query by product code) for some reasonable time frame. This data assists you in determining potential problem areas in the manufacture or design of the device, or lot or batch specific issues.

The establishment's reported registration and listing data should be verified during any GMP inspection to assure there have been no changes and the registration and listing data was accurately reported. Changes or inaccuracies should be immediately reported to the district medical device registration and listing monitor. See also Field Management Directive (FMD) 92.

510(k) and PMA data assists you in determining what devices the establishment is manufacturing and whether any new devices have been designed or changed since the last inspection. This data is useful in focusing the inspection on new or changed devices as well as devices that are higher risk devices, i.e., Class II or III versus Class I.

IOM 5.2 should be followed in regards to pre-announcement of medical device inspections.

5.6.2.2 - Quality Audit

The inspectional approach for identifying inadequate auditing of a quality assurance program is limited by the agency's policy, which prohibits access to audit results. The policy is stated in CPG section 130.300 (7151.02). Under the QS/GMP regulation (21 CFR 820.180 (c)) this prohibition extends to evaluations or audits of suppliers, 21 CFR 820.50(a), and Management Reviews conducted per 21 CFR 820.20. Evidence of inadequate auditing may be discovered without gaining access to the written audit reports. See the Guide to Inspections of Medical Device Manufacturers or Guide to Inspections of Quality Systems for inspectional guidance.

The preamble to the QS/GMP specifically states, "FDA will review the corrective and preventive action procedures and activities performed in conformance with those procedures without reviewing the internal audit reports. FDA wants to make it clear that corrective and preventive actions, to include the documentation of these activities, which result from internal audits and management reviews are not covered under the exemption at 820.180(c)." Therefore, these corrective and preventive actions and documentation are not excepted from inspectional scrutiny.

The QS/GMP regulation (21 CFR 820.180(c)) requires a manufacturer to certify in writing that audits and reaudits have been conducted whenever requested to do so by an investigator. Investigators through their supervisors should

consult with CDRH (HFZ-306) prior to requesting such
certification.

5.6.2.3 - Records

FDA has distinct authority under section 704(e) of the
FD&C Act [21 U.S.C. 374 (e)] to inspect and copy records
required under section 519 or 520(g) of the FD&C Act [21
U.S.C. 360i or 360j (g)]. Investigators should only collect
copies of documents as necessary to support observations
or to satisfy assignments. Manufacturers who have
petitioned for and obtained exemption from the QS/GMP
are not exempted from FDA authority to review and copy
complaints and records associated with investigation of
device failures and complaints.

You may advise manufacturers they may mark as
confidential those records they deem proprietary to aid FDA
in determining which information may be disclosed under
Freedom of Information.

Records must be maintained for as long as necessary to
facilitate evaluation of any report of adverse performance,
but not less than two years from the date the device is
released for distribution. Records required by the Radiation
Control for Health and Safety Act must be maintained for
five years. It is permissible to retain records in photocopy
form, providing the copies are true and accurate
reproductions.

5.6.2.4 - Complaint Files

Complaints are written or oral expressions of dissatisfaction
with finished device identity, quality, durability, reliability,
safety, effectiveness or performance. Routine requests for
service would not normally be considered complaints.
However, service requests should be reviewed to detect
complaints, and as part of any trend analysis system, and
to comply with 820.20(a)(3).

FDA has the authority to require a device firm to open its
complaint files, and review and copy documents from the
file.

Provisions in the FD&C Act pertaining to FDA review of
records are:

1. For restricted devices the FD&C Act in Section
    704(a)(1)(B) [21 U.S.C. 374 (a)(1)(B)] extends

inspection authority to records, files, papers, processes, controls and facilities bearing on restricted medical devices. See FD&C Act Sec. 704 [21 U.S.C. 374] for a full explanation and for a list of the items, e.g., financial data, which are exempt from disclosure to FDA.

2. For all devices, including restricted devices, refer to Section 704(e) of the FD&C Act [21 U.S.C. 374 (e)], which provides for access to, copying and verification of certain records.

3. Section 519 of the FD&C Act [21 U.S.C. 360i] requires manufacturers, importers, or distributors of devices intended for human use to maintain such records, and provide information as the Secretary may by Regulation reasonably require.

4. Section 520(g) of the FD&C Act [21 U.S.C. 360j (g)] covers the establishment of exemptions for devices for investigational use and the records which must be maintained and open for inspection.

QS/GMP requirements for complaint files are found in 21 CFR 820.198. GMP requirements for complaint files first became effective on December 18, 1978. The Quality System Regulation, which went into effect on June 1, 1997, added to and modified the requirements for complaint handling. The regulation contains a provision that records maintained in compliance with the QS/GMP must be available for review and copying by FDA (21 CFR 820.180). Complaint files are QS/GMP required records; therefore, the manufacturer must make all complaints received on or after December 18, 1978 and the records of their investigation available for FDA review and copying. EIRs should contain enough information to allow cross-referencing between complaints and MDRs.

21 CFR Part 803 requires medical device manufacturers to report deaths, serious illnesses, and serious injuries to FDA for which a device has or may have caused or contributed, and manufacturers must also report certain device malfunctions. The MDR reportable events must be maintained in a separate portion of the complaint files or otherwise clearly identified. These complaints must be investigated to determine whether the device failed to meet specifications; whether the device was being used for treatment or diagnosis; and the relationship, if any, of the device to the reported incident or adverse event.

When a firm determines complaint handling will be
conducted at a place other than the manufacturing site,
copies of the record of investigation of complaints must be
reasonably accessible at the actual manufacturing site.

5.6.3 - Sterile Devices

Inspections of sterile device manufacturers are conducted
per Compliance Program Guidance Manual 7382.845, as a
production process under the Production and Process
Control Subsystem. See the Guide to Inspections of Quality
Systems for further guidance.

5.6.4 - Labeling

Specific labeling requirements for in vitro diagnostics (IVDs)
are contained in 21 CFR 809.10.

Part 809.10(a) contains explicit labeling requirements for
the individual IVD containers, and for the outer package
labeling and/or kit labeling. Part 809.10(b) contains special
labeling requirements for the product insert, which must be
included with all IVD products. These two sections also
contain the requirements for: lot numbers, allowing
traceability to components (for reagents) or subassemblies
(for IVD instruments); stability studies for all forms of the
product; an expiration date, or other indication to assure the
product meets appropriate standards; and, the
requirements for establishing accuracy, precision,
specificity and sensitivity (as applicable).

Part 809.10(c) lists the labeling statements required for
IVDs which are being sold for investigational and research
use. Determine whether the firm is limiting the sale of IVDs,
labeled as such, to investigators or researchers. Document
any questionable products, and submit to CDRH for review.

Warning and caution statements recommended for certain
devices, along with certain restrictions for use, are
described in 21 CFR 801. This same section also contains
the general labeling regulations, which apply to all medical
devices.

5.6.5 - Government-Wide Quality Assurance Program (GWQAP)

Inspections under the GWQAP are conducted upon request
by OE, Division of Compliance & Information Quality

Assurance (HFC-240). Each assignment is specific and may involve more than a single compliance program. These inspections should be completed within 6 days from the date of the receipt from HFC-240. Specific questions arising during or as a result of these inspections should be directed to HFC-240.

5.6.6 - Contract Facilities

Device manufacturers may employ the services of outside laboratories, sterilization facilities, or other manufacturers (i.e., injection molders, packagers, etc). The finished device manufacturer is responsible for assuring these contractors comply with the QS/GMP and that the product or service provided is adequate. These contractors are subject to FDA inspection and some are subject to the QS/GMP regulation. This "...includes but is not limited to those who perform the functions of contract sterilization, installation, relabeling, remanufacturing, repacking, or specification development, and initial distributors of foreign entities performing these functions," per 21 CFR 820.3(o). Whether under contract or not if a firm manufactures a finished device by the definition found in 21 CFR 820.3(l) "Finished device means any device or accessory to any device that is suitable for use or capable of functioning, whether or not it is packaged, labeled, or sterilized they are subject to QS/GMP. NOTE: if the product manufactured by the contractor also meets the definition of a component and a finished device, the contractor is subject to the QS/GMP regulation.

Determine how a manufacturer evaluates and selects potential contractors for their ability to meet the manufacturer's requirements, as required by 820.50, Purchasing Controls. Conducting audits can be an effective method for assessment. However, not all contractors allow audits. Audits may not be feasible in some instances. In other instances the activity the contractor is conducting may not have a significant impact on the device safety or function; therefore, expending the resources necessary to audit the contractor may not be warranted.

Evaluations may be accomplished by other means such as requesting that the potential contractor fill out a questionnaire about their quality system, asking other customers of the contractor about their experiences with the firm, or basing assessments on past performance. Evaluations must be documented. The extent to which a manufacturer has evaluated a contractor, as well as the results of the evaluation, should govern the degree of oversight exercised over products and services supplied by the contractor.

5.6.7 - Small Manufacturers

When inspecting one-person or very small manufacturers for compliance with the QS/GMP master record and written procedure requirements, the investigator should realize that detailed written assembly, process, and other instructional procedures required for larger firms may not be needed. In a small firm, division of work is at a minimum, with one person often assembling and testing the finished device. In many cases, blueprints or engineering drawings could be adequate procedures. The QS regulation requires that certain activities be defined, documented and implemented. The regulation does not require separate procedures for each requirement and often several requirements can be met with a single procedure. The complexity of the procedures should be proportional to the complexity of the manufacturer's quality system, the complexity of the organizational structure and the complexity/risk of the finished device being produced. In assessing the need for detailed or lengthy written procedures, the investigator should make judgments based on training and experience of the individuals doing the work and the complexity of the manufacturing process. However, this does not mean small manufacturers have any less responsibility for complying with the QS regulation or assuring safe and effective devices are produced.

5.6.8 - Banned Devices

Section 516 of the FD&C Act [21 U.S.C. 360f] provides a device for human use may be banned by regulation (21 CFR 895) if it presents substantial deception or an unreasonable and substantial risk of illness or injury. Investigators should become familiar with this regulation. When you determine, during an inspection or investigation, that banned devices are being distributed, the distribution, manufacture, etc., should be documented as for any other violative product.

5.6.9 - Device Inspection Reports

See IOM 1.1. English language requirement. You should write your EIR following the guidance in IOM 5.10.4, 5.10.4.1, 5.10.4.2, 5.10.4.3. Section headings can be added to address the needs of other Compliance Program Guidance Manuals such as 7383.001 for pre-market and post-market PMA inspections. Include in your report the systems, processes, products, and product classification covered during the current inspection.

5.7 - *Biologics*

5.7.1 - Definition

A "biological product" means a virus, therapeutic serum, toxin, antitoxin, vaccine, blood, blood component or derivative, allergenic product, or analogous product, or arsphenamine or derivative of arsphenamine (or any other trivalent organic arsenic compound), applicable to the prevention, treatment, or cure of a disease or condition of human beings (Public Health Service Act Sec. 351(i)). Additional interpretation of the statutory language is found in 21 CFR 600.3. Biological products also meet the definition of either a drug or device under Sections 201(g) and (h) of the Federal Food, Drug, and Cosmetic Act (FD&C Act).

Veterinary biologicals are subject to the animal Virus, Serum, and Toxin Act which is enforced by USDA (21 U.S.C. 151-158).

5.7.2 - Biologics Inspections

FDA has developed a strategy known as "Team Biologics", a reinvention of the agency's approach to inspectional coverage of certain biological products. The periodic cGMP inspections and compliance operations of plasma fractionated products, allergenic products, vaccines, and biological in vitro diagnostic devices are now led by investigators and compliance officers in the Core Team. The Core Team investigators report to ORO headquarters; Core Team compliance officers report to OE. Inspections of certain CBER-regulated medical devices are not covered by the Core Team (e.g., blood establishment software) are conducted by District investigators who may or may not be part of the Biologics Cadre. See IOM 2.2 for a discussion of statutory authority. CBER maintains the lead for pre-licensing and pre-approval inspections of biological products, while ORA customarily leads PMA/510(k) inspections.

5.7.2.1 - Authority

Biological products are regulated under the authority of Section 351 of the Public Health Service Act and under the Food, Drug, and Cosmetic Act, as drugs or devices, with the exception of certain human cells, tissues, and cellular and tissue-based products (HCT/Ps) regulated solely under Section 361 of the Public Health Service Act (see 21 CFR 1271.10). Blood and blood products for transfusion are prescription drugs under the FD&C Act. Under the FD&C Act, source plasma and recovered plasma may have the legal identity of either a drug or device depending on its intended use. Section 351(a) of the PHS Act provides for licensure of biological products and inspection of the products covered is per 351(d). Most biological drugs are licensed. The investigational new drug application regulations (21 CFR 312) also apply to biological products subject to the licensing provisions of the PHS Act. However, investigations of blood grouping serum, reagent red blood cells, and anti-human globulin in-vitro diagnostic products may be exempted (21 CFR 312.2(b)).

5.7.2.1.1 - Blood and Source Plasma Inspections

The investigators in the Biologics Cadre perform inspections of blood and plasma establishments. For blood bank and source plasma establishment inspections (CP 7342.001 and 7342.002) use the CGMPs for Blood and Blood Components (21 CFR 606) as well as the general requirements for biological products (21 CFR Part 600), the general biological product standards (21 CFR Part 610), and the additional standards for human blood and blood products (21 CFR Part 640.) The drug GMPs (21 CFR 210/211) also apply to biological drugs. In the event it is impossible to comply with both sets of regulations, the regulation specifically applicable to the product applies. This would generally be Parts 606 and 640 of the regulations in the case of blood banks and source plasma establishments.

5.7.2.1.2 - Human Tissue Inspections

In February 1997, FDA proposed a new, comprehensive approach to the regulation of human cellular and tissue-based products (now called human cells, tissues, and cellular and tissue-based products or HCT/Ps). The agency announced its plans in two documents entitled, "Reinventing the Regulation of Human Tissue" and "A Proposed Approach to the Regulation of Cellular and Tissue-based Products" (62 FR 9721, March 4, 1997).

Since that time, the agency has published three final rules and one interim final rule to fully implement the proposed approach. On January 19, 2001, FDA finalized regulations to create a new, unified system for registering HCT/P establishments and for listing their HCT/Ps (registration final rule, 66 FR 5447). Part of the definition of "human cells, tissues, or cellular or tissue-based products" became effective on January 21, 2004. On January 27, 2004 (69 FR 3823), we issued an interim final rule to except human dura mater and human heart valve allografts from the scope of that definition until all of the tissue rules became final. On May 25, 2004, FDA finalized regulations requiring most cell and tissue donors to be tested and screened for relevant communicable diseases (donor-eligibility final rule, 69 FR 29786). On November 21, 2004, FDA finalized regulations requiring HCT/P establishments to follow current good tissue practice (CGTP), which governs the methods used in, and the facilities and controls used for, the manufacture of HCT/Ps; recordkeeping; and the establishment of a quality program. The new CGTP regulations also contain certain labeling and reporting requirements, as well as inspection and enforcement provisions (GTP final rule, 69 FR 68612). The donor eligibility and CGTP rules became effective May 25, 2005.

Part 1271 contains six subparts:

1. Subpart A of part 1271 - general provisions

2. Subpart B of part 1271 - registration

3. Subpart C of part 1271 - screening and testing of donors to determine eligibility

4. Subpart D of part 1271 - provisions on CGTP

5. Subpart E of part 1271 - certain labeling and reporting requirements

6. Subpart F of part 1271 - inspection and enforcement provisions.

The subparts apply as follows:

Subparts A through D apply to all HCT/Ps, i.e., to those HCT/Ps described in Sec. 1271.10 and regulated solely under section 361 of the PHS Act, and to those regulated as drugs, devices, and/or biological products. Subparts E and F, which pertain to labeling, reporting, inspection, and enforcement, apply only to those HCT/Ps described in Sec. 1271.10 and regulated solely under section 361 of the PHS Act. However, with the exception of two provisions (Sec. 1271.150(c) and 1271.155) subparts D and E are not being implemented for reproductive HCT/Ps described in 21 CFR 1271.10 and regulated solely under section 361 of the PHS Act.

HCT/Ps subject to the provisions of 21 CFR Part 1271 include, but are not limited to, bone, ligaments, skin, dura mater, heart valve, cornea, hematopoietic stem/progenitor cells derived from peripheral and cord blood, manipulated autologous chondrocytes, epithelial cells on a synthetic matrix, and semen or other reproductive tissue.

For HCT/P inspections, use the CPGM 7341.002, "Inspections of Human Cells, Tissues, and Cellular and Tissue-Based Products."

5.7.2.2 - Donor Confidentiality

Blood bank, source plasma, and human tissue establishments are sensitive to maintaining confidentiality of donor names. The mere reluctance to provide records is not a refusal. However, FDA has the authority under both the PHS and the FD&C Acts to make inspections and 21 CFR 600.22(g) and 1271.400(d) provides for copying records during a blood establishment inspection. For prescription drugs, section 704 of the FD&C Act specifically identifies records, files, papers, processes, controls, and facilities as being subject to inspection.

If you encounter problems accessing records, explain FDA's authority to copy these records. IOM 5.2.5 should be followed if a refusal is encountered. When donor names or other identifiers are necessary, they may be copied, but the information must be protected from inappropriate release. See IOM 5.3.8.6.

5.7.2.3 - Inspectional Objectives

The inspectional objective for biological products is to assure the products are safe, effective, and contain the quality and purity they purport to possess, and are properly labeled. The inspectional objective for HCT/Ps is to assure that HCT/Ps are recovered, processed, stored, labeled, packaged and distributed, and the donors are screened and tested, in a way that prevents the introduction, transmission, or spread of communicable diseases. Facilities will be inspected for conformance with:

1. Provisions of the PHS Act and FD&C Act,

2. Applicable regulations in 21 CFR 210-211, 600-680, and 820.

3. HCT/P regulations in 21 CFR 1270 and 1271.

4. FDA Policies, which include guidance to the industry, and the Compliance Policy Guides Chapter 2.

5.7.2.4 - Preparation

Review the district files of the facility to be inspected and familiarize yourself with its operation and compliance history. Review:

1. Appropriate Compliance Programs and related Compliance Policy Guides (CPG), Chapter 2.

   NOTE: Federal Cooperative Agreements Manual; MOU with the Department of Defense, and MOU with the Centers for Medicare and Medicaid Services (CMS) on transfusion services;

2. Correspondence from the firm depicting any changes since the last inspection;

3. Firm's registration and product listing information;

4. DFI's Guide to Inspections of Source Plasma Establishments, Guide to Inspections of Blood Banks,

and Guide to Inspections of Infectious Disease Marker
Testing Facilities.

5. Biological Product Deviation Reports, Adverse Reaction
    Reports, complaints, and recalls;

6. Guideline for Quality Assurance in Blood
    Establishments, (July 14, 1995).

Through guidance documents, CBER sets forth its
inspection policy and regulatory approach. A list of these
documents is attached to the current Compliance Program
Guidance Manuals (CPGM) available on the CBER internet
site at (http://www.fda.gov/cber/cpg/cpg.htm).

The OSHA regulation 29 CFR 1910.1030 dated December
6, 1991, was intended to protect health care workers from
blood borne pathogens, including those involved in the
collection and processing of blood products. The regulation
defines expectations for the use of gloves, hand washing
facilities, decontamination of work areas, waste containers,
labeling and training of employees and exemptions for
volunteer blood donor centers. FDA Investigators should
adhere to these safety guidelines during inspections or
related activities in establishments that process biologically
hazardous materials.

Become familiar with the OSHA regulations and their
applicability to 21 CFR 606.40(d)(1) and (2), which require
the safe and sanitary disposal for trash, items used in the
collection and processing of blood and for blood products
not suitable for use. Consult your district biologics monitor
for copies of the above references. Additional copies may
be obtained from ORO, Division of Field Investigations
(DFI), Biologics Group, HFC-130, 301-827-5653 or see
CBER's web site at http://www.fda.gov/cber.

5.7.2.5 - Inspectional Approach

Use the Compliance Program Guidance Manuals (CPGM)
and Guides to Inspection of Blood Banks, Source Plasma
Centers and Infectious Disease Marker Testing Facilities for
inspectional instructions. The EIR must clearly identify the
areas covered. The report should include a summary of the
inspection, the FDA 482, the FDA 483, if issued, and the
required FACTS EI Record.

Particular attention should be given to biological products
deviation reports indicative of problematic areas or

processes, adverse reactions, transfusion associated AIDS (TAA), transfusion or donation associated fatalities and hepatitis and HIV lookback procedures. For additional information regarding TAA, see CP 7342.001. The follow-up investigations to such reports should also be covered.

Complaints, in particular those involving criminal activity, must be promptly investigated and coordinated with other agency components as needed.

For blood banks and source plasma establishments, refer to CPGM 7342.001 and 7342.002 for a discussion of the systems approach to inspection. The CPGM incorporates a systems-based approach to conducting an inspection and identifies five (5) systems in a blood bank and source plasma establishment operation for inspection. Each system may not be in a particular establishment operation; therefore, the inspection should focus on the systems present. The CPGM directs an in-depth audit of the critical areas in each system. A multi-layered system of safeguards has been built into the blood collection, manufacturing and distribution system to assure a safe blood supply.

For HCT/P establishments, refer to CPGM 7341.002.

For Biological Drug Products, refer to CPGM 7345.848.

For Licensed Viral Marker Test Kits, refer to CPGM 7342.008.

If Investigators encounter products not specifically referenced in the regulations, they should contact CBER/OCBQ/ Division of Inspections and Surveillance for guidance.

5.7.2.6 - Regulations, Guidelines, Recommendations

Guidance documents for industry are made available to the public in accordance with good guidance practice regulations at 21 CFR 10.115. The contents of most of these documents are incorporated into the establishment's SOPs and/or license applications or supplements. Also, DFI has issued Blood Bank, Source Plasma Establishment and Infectious Disease Marker Testing Facility Inspectional Guides to be used by investigators during inspections.

Deviations from guidance documents must not be referenced on a FDA 483. However, since these documents are often related to specific GMP requirements, in most cases deviations can be referenced back to the GMP. If a deviation is observed during an inspection and the investigator relates it to the regulations or law, then the item may be reported on the FDA 483. During the discussion with management, the relationship of the deviation to the regulation or law, or accepted standard of industry, should be clearly explained.

If an establishment indicates it is not aware of any of these documents, provide the web site and the telephone number of CBER's Office of Communication, Training, and Manufacturers Assistance, 301-827-2000.

If a firm claims approval for an alternative procedure, verify by reviewing the firm's written approval letter. Approved alternative procedures may be verified by contacting CBER/Division of Blood Applications or the appropriate CBER product office.

5.7.2.7 - Technical Assistance

Several FDA regions and districts have biologics specialists who are available for technical assistance and consultation. Do not hesitate to avail yourself of their services.

The services of expert investigators in ORA/ORO/ Division of Field Investigations (DFI), Biologics Group, HFC-130, 301-827-5653, are available for telephone or on-site consultation and assistance in problem areas.

CBER/OCBQ, Division of Inspections and Surveillance (HFM-650), 301-827-6220, can provide technical assistance, and can coordinate assistance with other CBER offices.

5.7.2.8 - CBER Bio-research Monitoring

Bio-research monitoring (BIMO) assignments for biological products will generally be issued by the Center for Biologics Evaluation and Research (CBER) (see IOM 5.5.6).

5.7.3 - Registration, Listing and Licensing

5.7.3.1 - Registration and Listing

See IOM 2.9.3.1.

5.7.3.1.1 - Transfusion Services

Most transfusion services are exempt from registration under 21 CFR 607. This includes facilities that are certified under the Clinical Laboratory Improvement Amendments of 1988 (42 U.S.C. 263a) and 42 CFR Part 493 to perform the FDA-required tests on blood or has met equivalent requirements as determined by the Centers for Medicare and Medicaid Services, and are engaged in the compatibility testing and transfusion of blood and blood components, but which neither routinely collect nor process blood and blood components. Such facilities include establishments:

1. Collecting, processing and shipping blood and blood components under documented emergency situations,

2. Performing therapeutic phlebotomy and therapeutic plasma exchange after which the product is discarded,

3. Preparing recovered human plasma and red blood cells,

4. Pooling products/platelets for in-house transfusion,

5. Thawing frozen plasma or cryoprecipitate for transfusion.

5.7.3.1.2 - HCT/Ps

Establishments manufacturing HCT/Ps (human cells, tissues, or cellular or tissue-based products) as defined in 21 CFR 1271.3(d) must register and list using form FDA 3356. Examples of HCT/Ps include, but are not limited to, bone, ligament, skin, cornea, hematopoietic stem cells derived from peripheral and cord blood, manipulated autologous chondrocytes, and semen or other reproductive tissue. Establishments manufacturing HCT/Ps regulated as medical devices, drugs or biological drugs must also register and list with the FDA pursuant to 21 CFR 1271 using form FDA 3356.

5.7.3.1.3 - Laboratories

Laboratories performing infectious disease testing of donors of blood or blood components or HCT/P are an FDA obligation and required to register. Clinical laboratories

were previously exempted from registration by 21 CFR
607.65(g), but FDA revoked this regulation. Your
inspections should focus on activities relevant to blood
product and HCT/P testing operations.

5.7.3.1.4 - Military Blood Banks

Inspection of military blood banks is an ORA responsibility.
These facilities are required to meet the same standards as
other blood banks although military emergencies may
require deviations from the standards. A separate license is
held by each branch of the service; although each individual
establishment may be licensed or unlicensed, all are
required to register. Districts should notify the appropriate
military liaisons 30 days before inspection of a military
facility. For additional information on inspection of
government establishments, see Compliance Program
Guidance Manual 7342.001, the Federal Cooperative
Agreements Manual, and the MOU with Department of
Defense Regarding Licensure of Military Blood Banks.

Field Management Directive 92, Agency Establishment
Registration and Control Procedures, details the registration
process within the agency. Refer to FDA Compliance Policy
Guides (CPG), Chapter 2, Subchapter 230 (230.110), for
additional information on registration.

Ensure the firm's current registration forms reflect actual
operations.

5.7.3.2 - MOUs

Under the 1983 Memorandum of Understanding (MOU)
between the FDA and the Centers for Medicare and
Medicaid Services (CMS, formerly Health Care Financing
Administration - HCFA), CMS agreed to survey these
facilities that engage in minimal manufacturing in order to
minimize duplication of effort and reduce the burden on the
affected facilities while continuing to protect transfusion
recipients. However, no transfer of statutory functions or
authority is made under the MOU and the FDA retains legal
authority to inspect these unregistered transfusion services
whenever warranted. When appropriate, Districts should
conduct inspections jointly with the CMS regional liaison. If
you determine during a routine inspection an establishment
is a CMS obligation under the MOU, you should terminate
the inspection and report as such. See Federal Cooperative

Agreements Manual - FDA/HCFA Memorandum of
Understanding.

5.7.3.3 - Biologic License

See IOM 2.9.3.2. A biologics license application (BLA) shall
be approved only after inspection of the establishment(s)
listed in the application and upon a determination that the
establishment complies with the standards established in
the BLA and the requirements prescribed in applicable
regulations (21 CFR 601.20(d)). CBER is responsible for
conducting all pre-license (PLI) and pre-approval (PAI)
inspections of CBER-regulated products. These
inspections are part of the review of a BLA or BLA
supplement. CBER identifies the scope of the inspection
and invites ORA to participate in the inspections. Copies of
CBER's PLI and PAI inspection reports are forwarded to the
districts and should be part of the firm's file.

5.7.3.4 - Approval of Biological Devices

There must be a pre-approval inspection (PAI) of the
establishment for compliance with the QS/GMP regulation
and the firm's PMA. For licensed devices, CBER conducts
the pre-license inspection (PLI). Devices used in the
collection and testing of blood for transfusion are
approved/cleared through the PMA/510(k) authorities. ORA
Investigators customarily inspect the CBER regulated
devices, which are subject to PMA/510(k) applications.

5.7.4 - Responsible Individuals

In licensed establishments, the applicant or license holder
may designate an authorized official(s) to represent the
applicant to the FDA in matters of compliance. The FDA
482 and any 483 should be issued to the most responsible
person on the premises at the time of inspection. An exact
copy of the FDA 483 should also be forwarded to the top
official of the firm if that person did not receive the FDA
483. The designation as authorized official does not
necessarily mean that individual is the most responsible for
any non-compliance of the firm. In licensed or unlicensed
facilities, establish and document all individuals responsible
for violations and their reporting structure in the
organization.

5.7.5 - Testing Laboratories

Blood bank, source plasma, and HCT/P establishments may use outside testing laboratories to perform required testing.

Laboratories conducting testing for licensed blood banks are usually licensed. CBER may approve the use of a non-licensed laboratory to do required testing, provided the lab is capable of performing the tests and the lab registers with CBER prior to CBER approving the licensing arrangement.

Laboratories performing required testing for source plasma manufacturers must either be:

1. Licensed or

2. Certified to perform such testing on human specimens under the Clinical Laboratory Improvement Amendments of 1988 (42 U.S.C. 263a) and 42 CFR part 493, or has met equivalent requirements as determined by CMS.

Laboratories performing required testing for HCT/Ps must:

1. Test using approved FDA-licensed, approved or cleared donor screening tests according to the manufacturers instructions, and

2. be either certified to perform such testing on human specimens under the Clinical Laboratory Improvement Amendments of 1988 (42 U.S.C. 263a) and 42 CFR part 493, or has met equivalent requirements as determined by CMS.

Instructions for inspecting testing laboratories are included in the appropriate Compliance Program Guidance Manuals. Coordinate the inspection of non-registered laboratories with CMS regional office contacts. If a testing laboratory is located outside of the district, request an inspection by the appropriate district office, where appropriate.

5.7.6 - Brokers

Blood establishments may use brokers to locate buyers for products such as recovered plasma or expired red blood cells. These articles are used for further manufacture into products such as clinical chemistry controls and in-vitro diagnostic products not subject to licensure. Fractionators also use brokers to locate suppliers of plasma under the short supply provisions (21 CFR 601.22). During inspections, determine if the facility is selling products to any brokers. If brokers are used, determine if the brokered products are shipped to a facility operated by the broker or directly to the consignee.

Brokers who take physical possession of blood products and engage in activities considered manufacturing or labeling are required to register and are included in the OEI for routine inspection under the blood bank compliance program. Brokers who only arrange sales of or store blood and blood components, but do not engage in manufacturing activities are not required to register.

5.8 - Pesticides

5.8.1 - Pesticide Inspections

The objective of a Pesticide Inspection is to determine the likelihood of excessive residues of significant pesticides in or on products in consumer channels, and to develop sources of information for uncovering improper use of pesticide chemicals.

This requires directing coverage to two major areas:

1. Pesticide practices in the production and processing of field crops.

2. Application of pesticide chemicals in establishments storing and processing raw agricultural products.

Pesticide coverage must be provided during all food establishment inspections. Coverage of raw agricultural products will generally be on a growing-area basis.

Problem areas include:

1. *Improper use of pesticides around animals* - gross misuse of sprays and dips in animal husbandry may result in pesticide residues in foods.

2. *Use of contaminated animal feeds* - waste and spent materials from processing operations may contain heavy concentrations of pesticide residues, which were present in the original commodity. See Compliance Policy Guide 575.100.

3. *Past pesticide usage* - past pesticide practices on growing fields. Past use of persistent pesticides may result in excessive residues in the current food crop. You may need to check on pesticide usage for several years prior to an incident to ensure you gather enough information. Some pesticides last for many years in the environment.

5.8.2 - Current Practices

Cooperative Activities - important sources of information relative to evaluating the "Pesticide Environment" include:

1. At the start of the growing season, spray schedules recommended for each crop by county agents, state experiment stations, large pesticide dealers, farmers cooperatives, et al should be obtained.

2. Visits to agricultural advisors may provide information relative to heavy infestation of insect pests and fungal infections on specific crops in specific areas.

3. Daily radio broadcasts in most agricultural areas may provide information on spray schedules, insect pests, harvesting and shipping locations, etc.

4. Field employees of fruit and vegetable canning and freezing plants usually recommend spray schedules, pesticides, and harvesting schedules for products produced by contract growers.

5. United States Weather Bureau Offices and their reports will provide data on weather conditions, which may effect insect growth and their development, size of fruit or leaf growth, and dissipation of pesticide chemicals.

6. USDA Market News Service daily price quotations, and weekly quotations in trade magazines provide information regarding harvesting schedules since market prices are indicators of how quickly a crop will be harvested in a given area. Growers who have the opportunity to obtain high prices may harvest their crops without regard to recommended pre-harvest intervals.

7. State Colleges of Agriculture seminars or short courses on food and vegetable production may alert you to

significant departures from usual agricultural practices. Prior approval to attend such meetings should be secured from your supervisor.

8. Pesticide suppliers and distributors may provide information on spray practices, schedules, and the name and address of growers, etc.

NOTE: *The U.S. Department of Agriculture has a Pesticide Data Program (PDP), which provides data on pesticide use and residue detection. This program helps form the basis for conducting realistic dietary risk assessments and evaluating pesticide tolerances. Coordination of this program is multi-departmental, involving USDA, EPA and FDA, covered by a MOU (Federal Cooperative Agreements Manual). As a part of this program USDA collects data on agricultural chemical usage, and factors influencing chemical use, and collects pesticide residue data through cooperation with nine participating states. USDA provides this data to EPA, FDA and the public. Several USDA publications are listed below as reference material.*

The contact point at USDA for pesticide residue matters is:

Martha Lamont, Chief
Residue Branch, Science Division
Agricultural Marketing Service, USDA
8700 Centreville Road, Suite 200
Manassas, VA 221110
703-330-2300

Reference materials - the following reference materials provide background and data necessary or helpful in evaluating current practices. This material should be available at the District office.

1. Pesticide Chemicals - Regulations under the Federal Food, Drug and Cosmetic Act on tolerances for pesticides in food administered by the Environmental Protection Agency (EPA). (See 40 CFR 185)

2. EPA's Pesticide Regulations - Tolerances for Raw Agriculture Products. (See 40 CFR 180)

3. EPA's Rebuttable Presumption Against Registration (RPAR) List.

4. Pesticide Index. - By William J. Wiswesser. A publication containing information on trade names, composition and uses of commercial pesticide formulations.

5. The Daily Summary or Weekly Summary. News releases and reports from USDA.

6. USDA's Weekly Summary Shipments-Unloads.

7. Agricultural Economic Report No. 717 Pesticide and Fertilizer Use and Trends in U.S Agriculture (May 1995)

8. Annual Pesticide Data Summary

9. Reports from USDA's Crop Reporting Board.

10. USDA's Pesticide Assessment Reports.

5.8.3 - Growers

Preliminary investigation of growing areas at the start of the season will provide data necessary for district work planning including production schedules, types and acreage of crops, pesticides used and the names and addresses of growers and shippers.

Growing Dates - The significant growing dates relative to pesticide usage are as follows:

1. Planting date,

2. Date of full bloom, and

3. Date of edible parts formation.

Harvest Dates - The dates of the anticipated harvest season will provide planning information relative to pre-harvest application and shipping.

Acreage - This will provide volume information for work planning.

5.8.3.1 - Pesticide Application

Ascertain the actual pesticide application pattern for each crop. Look for objective evidence to document actual grower practice. Check the grower's supply of pesticide chemicals, look for used pesticide containers, visit his source of supply, etc. Check spraying and dusting practices. Establish if pesticide chemicals are used in such a manner that excessive residues might result.

The following information provides a basis for evaluating pesticide usage:

1. *Pesticide Chemical Applied* - List the common name if there is no doubt as to the chemical identity of the pesticide. Include labeling indications and instructions.

2. *Method of Application* - Describe the method of application i.e., ground rig, airplane, greenhouse aerosol, hand, etc.

3. *Formulation* - Describe the formulation i.e., wettable powder, emulsifiable concentrate, dust, granules, aerosol, etc. Express as pounds of active ingredient per gallon or percent wettable powder.

4. Number of Applications and Dates.

5. *Rate of Last Application* - Calculate the amount of active ingredient per acre.

6. *Pre-Harvest Interval (PHI)* - Calculate the number of days between the day of the last application of pesticide and the harvest date or anticipated harvest date. Compare to the PHI.

7. Visible residue on grower's crop.

8. *Summary of Usage* - Determine the USDA Summary Limitations and evaluate the responsible usage.

5.8.3.2 - Pesticide Misuse/Drift/Soil Contamination

Pesticide residues, which exceed established tolerances, action levels, or "regulatory analytical limits" may be caused by pesticide misuse which can include:

1. Excessive application of a chemical on a permitted crop.

2. Failure to follow labeled time intervals between the last pesticide application and harvest.

3. Use of a non-approved pesticide on a crop.

4. Failure to wash a crop when pesticide labeling requires it (e.g., for certain EBDC's).

Other conditions, which may cause illegal residues, include spray drift and soil contamination.

Drift may be documented by determining which crops and pesticides have been grown/used in fields adjacent to those sampled. Determine direction of prevailing winds and wind condition on the day of spraying. Selective sampling will aid in determining if drift occurred. Compliance Samples collected to document pesticide drift should be Flagged and noted in block 16 of the CR as "Drift Sample - Maintain as Individual Subs".

Soil contamination by compounds, which are relatively stable in the environment, may cause systemic uptake of the compounds by growing crops. Follow-up investigations to violative samples may, in some limited cases, include soil samples as an attempt to determine the source of the contaminant. Do not routinely collect soil samples.

5.8.4 - Packers and Shippers

Follow the same general procedure as in IOM 5.8.3. Observe and report the following:

1. *Treatment Before Shipping* - This may include stripping of leaves, washing, vacuum cooling, application of post-harvest preservative chemicals, use of cartons with mold-inhibiting chemicals, waxes, colors, fumigation, etc.

2. *Identification of Growers' Lots* - Determine procedure or methods used to maintain the identity of each grower's lot. Provide the code and key if any.

3. *Labeling* - Quote labeling or brand names.

4. *Responsibility* - Determine whether the packer or shipper knows what sprays have been used on the products shipped.

5.8.5 - Pesticide Suppliers

Pesticide suppliers should be visited routinely during growing-area coverage. They may provide valuable information about pesticides being used on various crops in the growing area. Some suppliers may suggest spray schedules or advise growers about pesticide usage.

Determine what representations were made by the manufacturer of pesticide chemicals for which there is only a temporary tolerance or experimental permit. Get copies of any correspondence relating to sale and use of these products. Obtain names of growers to whom sales are made if such sale was not for use on acreage assigned under the experimental permit. Collect Official Samples of any crops treated with the pesticide.

5.8.6 - Pesticide Applicators

Pesticide applicators may provide valuable information about pesticides being used on various crops in the growing area. Interview several pesticide applicators, particularly those using airborne equipment. Determine the pesticide chemicals, their formulation, and on what crops they are currently being applied. Determine who supplies the pesticides and how they are prepared to assure proper concentration. If state law requires the applicator to keep a record of each spray application, request permission to review such records. Determine what steps are taken to assure drift on adjoining crops does not result in violative residues. Where there is likelihood of drift, collect Selective Samples from adjoining fields.

5.8.7 - Sample Collections

See IOM Sample Schedule Chart 3 - Pesticides.

5.9 - Veterinary Medicine

5.9.1 - CVM Website

The Center for Veterinary Medicine has its own website. The website contains an alphabetical listing of topics under "Index"; a listing of current and planned Guidance Documents; and on line access to the "Green Book" database listing animal drug approvals. There is a "search" feature allowing you to search for documents containing various words or phrases. The website also contains organizational information for the Center and an explanation of the various laws and regulations which the Center enforces. Information on the website can provide guidance for inspectional efforts related to CVM obligations.

5.9.2 - Veterinary Drug Activities

CVM is responsible for inspections of therapeutic and production drugs, and Active Pharmaceutical Ingredients (APIs). Therapeutic drugs are used in the diagnosis, cure, mitigation, treatment or prevention of disease. Production drugs are used for economic enhancement of animal productivity. Examples include: growth promotion, feed efficiency and increased milk production.

Preapproval inspections are conducted pursuant to pending NADA or ANADA applications.

Post approval inspections of veterinary drugs are conducted to determine compliance with the Current Good Manufacturing Practices (CGMPs) for Finished Pharmaceuticals under 21 CFR Part 211. These cGMPs apply to both human and veterinary drugs. Information on veterinary drugs approved can be found in the "Green Book" database accessed through CVM's website.

APIs are active pharmaceutical ingredients. Many of the APIs used to manufacture dosage form drugs are imported from foreign countries. The intended source for an API must be indicated in NADA/ANADA submissions for new animal drug approvals. Any change in a source for an API would require a supplement to the application.

Extra label drug use refers to the regulations in 21 CFR Part 530 codified as a result of the Animal Medicinal Drug Use Clarification Act (AMDUCA) of 1994. These regulations set forth the requirements for veterinarians to prescribe extra label uses of certain approved animal and human drugs and the requirements for the existence of a valid veterinarian/client/patient relationship (VCPR). The regulations under 21 CFR Part 530 address issues regarding extra label use in non-food as well as food producing animals. 21 CFR 530.41 contains a list of drugs that cannot be used in an extra label fashion in food-producing animals. During an inspection or investigation if you encounter any situations on extra label use of the listed drugs, you should contact CVM's Division of Compliance (HFV-230) (240-276-9200).

The regulations under 21 CFR Part 530 also address compounding of products from approved animal or human drugs by a pharmacist or veterinarian. The regulations clearly state compounding is not permitted from bulk drugs. This would include APIs. CVM has an existing CPG on Compounding of Drugs for Use in Animals (CPG 608.400). A copy can be found on CVM's website. The Division of Compliance (HFV-230) has issued assignments to conduct inspections of firms, including internet pharmacies, who may be engaged in the practice of manufacturing under the guise of pharmacy compounding. You should contact the Division of Compliance (HFV-230) at 240-276-9200 to report instances of compounding or to seek guidance on inspectional issues, or regulatory and enforcement policies.

5.9.3 - Medicated Feeds and Type A Articles

Animal feed is defined under section 201(w) of the FD&C Act [21 U.S.C. 321 (w)]. CVM is responsible for control of medicated and non-medicated animal feeds, Type A medicated articles and pet foods.

The regulations for animal food labeling are in 21 CFR Part 501. The regulations for medicated feed mill licensure are in 21 CFR Part 515. The cGMPs for Medicated Feeds are in 21 CFR Part 225. The cGMPs for Type A Articles are in 21 CFR Part 226.

Inspections are routinely conducted of medicated feed mills and manufacturers of Type A Medicated Articles.

If you have questions related to cGMPs and enforcement policies and strategies concerning Medicated Feeds and Type A Articles you should contact the CVM/Division of Compliance (240-276-9200).

Guidance on pet food labeling requirements can be found on CVM's website.

5.9.4 - BSE Activities

CVM is responsible for FDA's educational and regulatory activities involving BSE. BSE is "Bovine Spongiform Encephalopathy" and is often referred to as "mad cow disease." BSE information can be found on the CVM website. CVM has four Guidance Documents in place dealing with BSE (67-70, dated February 1998). The Guidance Documents address renderers, protein blenders, feed manufacturers, distributors and on farm feeders.

Investigations Operations Manual Chapter 5:
Establishment Inspections

Questions on inspectional assignments and regulatory
activities in the BSE area should be addressed to the
CVM/Division of Compliance (HFV-230) at 240-276-9200.

5.9.5 - Tissue Residues

The presence of violative drug residues in food from
slaughtered animals is a human health concern. Tissue
residue investigations/inspections are performed in
response to reports of violative drug residue levels found in
tissue sampled at slaughter by the USDA.

Tissue residues are commonly caused by the medication of
animals prior to marketing and failure to follow the
withdrawal times. When a new animal drug is approved the
manufacturer must conduct studies to accurately determine
withdrawal times. Allowable tolerances for residues of new
animal drugs in food can be found in 21 CFR Part 556.

Tissue residue inspections and investigations are unique in
comparison to other fieldwork. Although your investigation
may begin at the establishment or person named on the
USDA/FSIS "Warning letter," you may inspect and/or visit
more sites as part of your overall investigation. You may
have to visit an auction barn, dealer, trucker, veterinarian,
drug supplier, slaughter facility (USDA firm management or
State personnel), etc. One or more of these establishments
may have caused or allowed the tissue residue to occur.
Thus, each establishment's activities may warrant a
recommendation for regulatory action such as Warning
Letter, Injunction, etc. should the establishment's
involvement with residue violations continue.

Upon receipt of a FACTS assignment from CVM to conduct
a tissue residue follow-up investigation, the district may
create a second assignment, linked to the original CVM
assignment, which will include all operations required to
complete the assignment. This could include multiple
inspections, sample collections and/or investigations. You
may not be aware of all the establishments prior to
beginning your investigation. Appropriate operations
should be added to or deleted from the district assignment.

Each site visit is unique and each produces its own set of
unique documents and evidence requiring individual
reporting by establishment. You should use good judgment
during case development to assure you document your
investigation thoroughly. Explain the chain of events and
evidence, from the initial tissue residue report, and how

other establishments were involved. Collect samples (usually DOC samples) as appropriate. Consultation with your supervisor and/or compliance branch during these operations is essential to assure all evidence necessary to develop a quality case is obtained and submitted in an appropriate format.

Following completion of all operations, you should prepare a Memo of Investigation referencing the FACTS assignments for your supervisor's endorsement to the district Compliance Branch, with a copy to the originating CVM office. This Memo will summarize each site visit (EI or Investigation), sample(s) collected and relevance to the overall CVM assignment. A copy of the memo will be routed to each appropriate factory file.

The individual operations will then stand alone and/or may be used together to build one or multiple cases.

For example, a site visit to a slaughter facility may obtain information on the animal from the USDA inspection personnel on site; and obtain verification from management the establishment ships in interstate commerce. Information obtained at the slaughter facility or other establishments may be documented in an affidavit from each individual providing salient information. A site visit to a veterinarian may be important to establish whether the drugs which caused the tissue residue(s) were prescribed and, if so, how they were prescribed. When there is reason to believe off-label use or other activities have occurred which may warrant a recommendation for regulatory action, an establishment inspection should be conducted and your evidence included with your report.

For information on tissue residue violations and activities you should contact the CVM/Division of Compliance (HFV-230, 240-276-9200).

5.9.6 - Veterinary Devices

Medical devices for animal/veterinary use are not subject to the premarket approval requirements like human medical devices. Once an animal use device is marketed the Center is concerned with safety and efficacy of the veterinary device. CVM often recommends firms use the human device GMPs in controlling the manufacturing of animal use devices. CVM also suggests labeling be sent in for review by the Division of Compliance (HFV-230) to avoid misbranding. Regulatory questions for veterinary/animal

use devices should be directed to the CVM/Division of
Compliance (HFV-230).

5.9.7 - Animal Grooming Aids

Grooming aids for animals formulated and labeled only to
cleanse or beautify the animal are not cosmetics within the
meaning of Section 201(i) and not subject to the Federal
Food, Drug, and Cosmetic Act. Where animal grooming
aids are labeled to contain an active drug ingredient or
otherwise suggest or imply therapeutic benefit, they may be
considered to be drugs and/or new animal drugs as defined
by Section 201(v) of the Act (see CPG 7125.21).

Questions on labeling and regulatory concerns should be
directed to the Division of Compliance (HFV-230) at 240-
276-9200.

5.9.8 - CVM Bio-Research Monitoring

CVM issues assignments to the field to conduct BIMO
inspections of animal drug studies, including both
therapeutic and production drugs. Currently, there is no
requirement for animal drug studies to be controlled by any
sort of institutional review board (IRB). See IOM 5.5.6.

5.10 - Reporting

5.10.1 - Establishment Inspection Report (EIR)

See IOM 1.1 English language requirement. The EIR
consists of the following in this order: a printed copy of the
FACTS Establishment Inspection Record (EI Record)
including, at least, the endorsement with the EIR
distribution printed at the bottom of the "endorsement"
section of the EI Record; carbon or other copies of FDA
forms issued during the inspection such as the FDA 482,
FDA 483, and FDA 484; investigator's narrative report; copy
of assignment if available; exhibits; and/or any additional
material attached and referred to in the narrative report.
Regarding the use of checklists (such as the BSE
Checklist), the original raw data completed checklist should
be submitted with the EIR. If you maintain the data in your
regulatory notes, rather than entering directly on the form,
then enter on the electronic copy. A printed copy from
FACTS becomes the data to include with the EIR.

No copies of inspection reports will be maintained other than in headquarters, district, and resident post files. The signed original report is maintained in the district office or in the case of foreign inspections in the appropriate Center HQ office.

5.10.2 - Endorsement

The endorsement of the establishment inspection is prepared by the supervisor. Some supervisors may have the investigator prepare proposed endorsements. Endorsements should fit in the available space provided in FACTS. If the endorsement exceeds the 2000 character space provided in FACTS, a separate endorsement should be prepared, fully identifying the firm with a Summary of the Endorsement included in FACTS. The FACTS EI Record will be printed and used as the endorsement and routing document to accompany the EIR. See also IOM 5.10.4.1.

Normally the endorsement consists of:

1. The reason for the EI, i.e., workplan, or assignments from headquarters. State the subject of the assignment and reference. If the assignment was issued hard copy (i.e. not through FACTS), it should be attached to the EIR following the narrative. Include the FACTS assignment number and compliance tracking number if applicable.

2. A brief history of previous findings including classification of previous EI, any action taken by the district and/or corrective action taken by the firm in response to inspectional observations from the previous inspection.

3. A concise summary and evaluation of current findings and samples collected.

4. Refusals, voluntary corrections or promises made by the firm's management.

5. Classification and follow-up consistent with inspectional findings and Agency policy including notification of other districts and headquarters as warranted.

6. Distribution consistent with District policy and the requirements of the specific Compliance Program Guidance Manual(s).

7. *Note:* route a copy of the FACTS Establishment Inspection Record and the EIR to DIOP (HFC-170) when any violative, imported products are identified.

8. Per CPG section 110.300, do not report the FURLS Registration number.

The existence of Personal Safety Alerts (IOM 5.2.1.3) or Situational Plans (IOM 5.2.1.4) pertaining to the firm should be included in the endorsement section only and not in the EIR. The signed endorsement should be updated to indicate if an addendum to the EIR (IOM 5.10.6) or an amended FDA 483 (IOM 5.2.3.1.6.2) has occurred.

PROFILES: Updating the Field Accomplishments and Compliance Tracking System (FACTS) database with a Compliance Status for each profile class code associated with the firm's operations and/or products, is the responsibility of Field and Center Investigators, Supervisors and Compliance Officers.

For Domestic inspections, hardcopy or e-mail notification of Potential OAIs are not necessary. FACTS automatically sends OAI Notifications to DCIQA (HFC-240) electronically.

For foreign inspections, the instructions in Exhibit 5-14, Updating Profile Data in FACTS-Guidance, states "When a potential OAI Notification cannot immediately be entered in the FACTS firm profile record, the investigator should notify the Division of Field Investigations (DFI) of the potential OAI situation via FAX (301-827-6685 or 301-443-6919) as soon as the potential OAI situation is known and during the investigation. DFI will notify the appropriate Center and Division of Compliance Information and Quality Assurance (DCIQA) at 240-632-6824 or by e-mail to ORA HQ DCIQA Employees."

See Exhibit 5-14. The COMSTAT Guidance to Field and Centers document can be accessed from the DCIQA web site.

5.10.2.1 - Compliance Achievement Reporting System (CARS)

FACTS is used to report achieved and verified compliance actions, which are not the result of a legal action. A compliance achievement is the observed repair, modification, or adjustment of a violative condition, or the repair, modification, adjustment, relabeling, or destruction of a violative product when either the product or condition does not comply with the Acts enforced by the FDA.

5.10.2.1.1 - Reporting Criteria

There are three criteria for reporting into the CARS system:

1. The detection or identification of the problem. A problem may be observed by FDA, other federal officials, or by state or local authorities and referred to FDA; and as a result of an inspection, investigation, sample analysis, or detention accomplished by ORA or states under contract to ORA.

2. The correction of the problem. The correction is directly attributable to the efforts of ORA or state officials under contract to ORA (involving contract products only); and is unrelated to the filing of a legal action, i.e., seizure, prosecution, injunction.

3. The verification of the correction of the problem. The correction is verified by the FDA, other federal officials or state or local authorities and reported in writing to the FDA; and is based on an inspection, investigation, sample analysis, or letter from a firm to FDA certifying the problem has been corrected.

5.10.2.1.2 - Data Elements

Only when the corrective action(s) has been verified should a CARS be reported. The data elements are those entered/coded in FACTS (See IOM Exhibit 5-15):

1. *PAC.* See the Data Codes Manual. Should there be insufficient space to code all corrections verified on an occasion, record the most significant corrections.

2. *PROBLEM TYPE.* The problem type is the problem(s) identified during the operation(s). Use the List of Values

(LOV) found in this field on the Compliance
Achievement Reporting Screen. If "Other" is chosen,
you should include an explanation in the 'Remarks"
field.

3. *CORRECTIVE ACTION.* The action the establishment
   took to correct the identified problem. Use the LOVs
   found in this field on the CARS screen. If "Other" is
   selected, you should include an explanation in the
   "Remarks" field.

4. *VERIFICATION DATE.* Use the date the corrective
   action(s) is verified, either through an establishment
   inspection, an investigation, or a letter from the
   establishment certifying the corrections have been
   made. Include documentation to verify the action such
   as repair receipts/plans.

5. *CORRECTING ORGANIZATION.* The FDA, other
   federal agency, or state or local authority, which
   observed the verified correction. Use the LOVs found in
   this field on the CARS screen.

6. *REPORTING DISTRICT.* The FDA, other federal
   agency, or state or local authority, which is actually
   inputting the verified correction. Use the LOVs found in
   this field on the CARS screen.

7. *REASON FOR CORRECTION.* The action the FDA
   took to make the correction happen. Use the LOVs
   found in this field on the CARS screen. If "Other" is
   chosen, you should include an explanation in the
   "Remarks" field.

5.10.3 - Facts Establishment Inspection Record (EI Record)

Per FMD-130, each ORA District is responsible to ensure
all investigators verify, correct, and enter changes to the
OEI (including Profile data for profilable firms) on the firm's
maintenance screens in FACTS during each inspection,
investigation and during any OEI update. Consult with your
supervisor and District OEI Coordinator to assure data is
accurately updated. See IOM Exhibits 5-15 and 5-16. The
FACTS Profile Data instructions and FACTS generated
assignment are attached as IOM Exhibits 5-9 and 5-14.

Inspectional accountable time in FACTS consists of the hours devoted to file reviews (operational preparation), actual inspectional, investigational, audit, etc. time (onsite), document (exhibit) preparation and EIR (report) write-up. Accountable time does not include travel time. One occasional exception could be when more than one participant in an inspection/investigation travel together and discuss/prepare while in route.

5.10.4 - Narrative Report

See IOM 1.1 English language requirement. You should use Turbo EIR for all EIRs. The narrative report is the written portion of the EIR, which accurately describes the investigator's inspectional findings. The narrative report may be prepared in two formats depending on the type of inspection, inspection classification, and program area. A Summary of Findings narrative report is used for non-violative, non-initial inspections - see IOM 5.10.4.1. The full Standard narrative report is used for human drug and medical device inspections, initial and potential Official Action Indicated (OAI) classified inspections in other program areas - see IOM 5.10.4.3. The "Summary of Findings" report format may be used for some Voluntary Action Indicated (VAI) classified inspections as directed by your supervisor. Additional requirements for human drug and medical device reports are described in IOM 5.5.8 and 5.6.9. For all reporting formats, include additional information as directed by your assignment, Compliance Program Guidance Manual, or your Supervisor.

All reports should be prepared as stand-alone documents outside of FACTS. Your Establishment Inspection Report (EIR) should:

1. Be factual, objective, and free of unsupportable conclusions.

2. Be concise while covering the necessary aspects of the inspection.

3. Not include opinions about administrative or regulatory follow-up.

4. Be written in the first person.

5. Be signed by all FDA and commissioned personnel participating in the inspection. See IOM section 5.1.2.5.1 when more than one FDA or commissioned person participated in the inspection.

5.10.4.1 - Non-Violative Establishments

Investigators should use "Summary of Findings", stand-alone, narrative reports for non-violative establishments, unless otherwise directed by your supervisor, the assignment or the Compliance Program Guidance Manual involved.

Exception: human drug and medical device GMP inspection reports, which have additional reporting requirements should be written in the Standard Narrative report format as in 5.10.4.3.

The Summary of Findings Report may not be written solely in the FACTS provided "Inspection Summary" heading. The Summary of Findings report should include:

1. The reason for the inspection;

2. The date, classification and findings of the previous inspection;

3. The actual inclusive dates of the inspection (these may be included as part of a header or in the body of the EIR.)

4. The name of the person to whom credentials were shown and the Notice of Inspection was issued and the person's authority to receive the Notice. Explain if you were unable to show credentials or issue forms to top management;

5. The scope of the inspection; i.e., comprehensive or directed; and a brief description of the products, processes or systems covered during the inspection; the manufacturing codes and if necessary their interpretation.

6. The significant findings if any;

7. Management's response or corrections;

8. Warnings given to management; and

9. The investigator's handwritten signature.

5.10.4.2 - Violative Establishments

For domestic inspections where regulatory action is being recommended and when the District has final classification responsibility, the inspection report should normally be submitted within 10 days to the District or Center Compliance Branch as per established procedures. Please note, that depending on the type and severity of the regulatory action, it may be necessary to submit the EIR in less than 10 days. You should consult with your supervisory investigator in these instances. Refer to FMD-86 and the Regulatory Procedures Manual regarding other timeframes associated with non-violative inspections.

All violative EIR's should in addition to the information required for non-violative reports contain the following:

1. The objectionable conditions or practices described in sufficient detail so someone reading the report will clearly understand the observation(s) and significance.

2. The objectionable conditions or practices cross-referenced to FDA 483 citations, samples collected, photographs, or other documentation including exhibits attached to the EIR.

3. Information as to when the objectionable conditions or practices occurred, why they occurred, and who is or was responsible, developed to the highest level in the firm.

5.10.4.3 - Individual Narrative Headings

There are many acceptable ways of organizing a narrative report. The key is to cover the required information in IOM 5.10.4 and 5.10.4.2, or as required by the assignment, Compliance Program Guidance Manual, or your supervisor. The appropriate use of headings should not result in repetition of the same information in different sections. You are encouraged to create headings as necessary to present the inspectional findings in the most concise manner. For non-violative and some VAI reports, a single heading such as "Summary of Findings" is sufficient (for exceptions, see IOM 5.10.4.1). Turbo EIR should be used to generate the FDA 483. In certain instances, if you experience computer problems, do not delay the issuance of the FDA 483. See IOM 5.2.3. You should use Turbo EIR for all EIRs.

5.10.4.3.1 - *STANDARD NARRATIVE REPORT*

STANDARD NARRATIVE REPORT: HEADINGS, CONTENT AND ARRANGEMENT OF YOUR REPORT

Use the Standard narrative report format for all program areas. The Standard narrative format contains sections within specific headings. Reporting requirements under these headings fall into two categories: those which should be reported every time (if applicable) and those which only need to be reported if an element has changed.

Initial or potential OAI classified inspections: complete a full standard narrative report for all program areas. You should include a Table of Contents for all complex or full standard narrative reports.

Note: *All human drug and medical device inspection reports should be full narrative Reports. You should add the supplemental information listed under the subheadings for human drug and medical device inspection reports as appropriate. Human drug inspection reports do not need full and detailed narratives for every heading. In many cases, brief summaries addressing the format areas will be sufficient.*

5.10.4.3.2 - *Summary*

Summary:

1. Provide the reason for the inspection (e.g., compliance program, by assignment, etc.);

2. The scope of the inspection (comprehensive, directed, sample collection only, QSIT level, etc.).

3. Provide a summary of the findings, date, and classification of the previous inspection and the firm's response/corrective actions.

4. List the products, systems and processes covered during the current inspection, and the types of records and documents reviewed. For human drug reports, list the systems not covered.

5. Provide a summary of the current findings, refusals, samples collected, warnings given to management, and a summary of management's response or voluntary corrections.

6. Per CPG section 110.300, do not report the FURLS Registration number.

5.10.4.3.3 - Administrative Data

Administrative Data:

1. The firm name, address, phone, FAX and e-mail address.

2. Report the names and titles of the Investigator(s), Analyst(s), non-FDA officials, etc. Report the name of the firm's responsible official who gave permission to non-FDA officials without inspection authority to accompany you during your inspection. See IOM 5.1.1 and 5.2.2.

3. The inclusive date(s) of the current inspection, i.e., list the actual dates in the plant.

4. If a team inspection and some individuals were not present during the entire inspection, indicate dates in plant for each team member.

Report Full Names and Titles of:

1. To whom FDA Official Credentials were shown,

2. To whom any FDA forms were issued to or signed by during the inspection (FDA 482, 483, 484, 463, etc.); where appropriate, explain the reason a form(s) was not issued to or signed by the most responsible individual (this may be reported in the Individual Responsibility and Persons Interviewed heading below),

3. Who wrote which section of the EIR, if this was a team inspection report, and

4. In-plant inspectors or other government agencies (IOM 5.4.9).

5.10.4.3.4 - History

History:

1. Report the legal status of the firm (corporation, partnership, limited liability corporation, etc.). If a corporation, list in which state and when the firm was incorporated.

2. List the parent corporation, corporate address and any subsidiaries.

3. Provide a summary of any regulatory actions and prior warnings (do not cite any action only recommended but not approved). You should also report any significant/relevant inspectional history pertinent to the current EI or recommendation.

4. Include any relevant recalls, etc. since the last inspection.

5. Report the hours of operation and any changes from past inspections (include seasonal variations).

6. Report the current registration(s) status or any changes to registration status. Per CPG section 110.300, do not report the FURLS Registration number.

7. If directions to the firm would be helpful in future visits, include the information.

8. Provide the names, titles and addresses of top management official(s) to whom correspondence should be addressed (FMD 145, W/L, etc.).

9. For foreign inspections, list U.S. consignees to whom the firm's products are shipped.

10. For Human Drugs - domestic firms, identify the general types of customers and provide the names and addresses for several regular customers of a few of the firm's products.

5.10.4.3.5 - Interstate (I.S.) Commerce

Interstate (I.S.) Commerce:

1. Report changes in the previous estimate of the percentage of products shipped outside of the state (or exported to the U.S.) and the basis of the estimate.

2. Report the firm's general promotion and distribution patterns.

3. If there is an apparent violative product, provide examples of I.S. shipments of violative product(s); or

4. If no such shipments, provide examples of I.S. shipments of major components of apparent violative products - with complete I.S. documentation in either case.

5.10.4.3.6 - Jurisdiction (Products Manufactured and/or Distributed)

Jurisdiction (Products Manufactured and/or Distributed):

1. Include a list of a representative number of currently marketed products subject to FD&C Act or other statute enforced by FDA or counterpart state agency, including any believed violative.

2. Collect appropriate labeling (product and case labels, inserts, brochures, manuals, promotional materials of any type) for those products believed violative or representing any significant new or unusual operation, industry or technology; or as directed by your supervisor.

3. Document any applicable labeling agreements (and obtain a copy) and statutory guaranty given or received per Sections 301(h) and 303(c)(2) of the FD&C Act [21 U.S.C. 321 (h) and 333 (c)(2)] (IOM 5.3.7.2)

In addition, the label, labeling and promotional materials are a critical part of determining a product's intended use.

1. In instances where a regulatory action is being considered based on product labels, labeling, and/or other promotional materials, including any Internet websites, you should collect all available documentation. This includes all written, printed or graphic matter on the immediate container of an article or accompanying the article (the product's label and

labeling, see FD&C Act, 201(k) and (m) [21 U.S.C. 321(k) and (m)] and IOM 4.4.9.1). Accompanying labeling could include brochures, pamphlets, circulars, and flyers, as well as audio and video tapes.

2. In cases where there may be a dispute about whether a product is a drug or a dietary supplement, you should collect all materials which claim a product can be used for the treatment of any disease.

5.10.4.3.7 - Individual Responsibility and Persons Interviewed

Report with whom you dealt, and in what regard (both during and prior to the start of the inspection):

1. Who provided relevant information,

2. Who accompanied you during the inspection,

3. Who refused access to required records or any other refusal of information (Note: a separate heading for Refusals may be needed if refusals are significant, extensive or an Inspection Warrant is anticipated),

4. Who refused to permit inspection (IOM 5.2.5.1) and

5. For Human Drug inspection reports, also include the name, title, physical mailing address, phone, and fax number and e-mail address for any U.S. Agent or broker who represents the company when dealing with the FDA.

Describe roles and authorities of responsible individuals, including the full names and titles of individuals providing you with information.

Report changes to the following:

1. Who is the most responsible individual at the inspected firm? Who is the responsible head or designated correspondent? Refer to IOM 5.3.6, 5.3.6.1, and 5.3.6.2.

2. Report full names and titles of owners, partners, and corporate officers. Who has the duty, power and responsibility, and authority to prevent, detect, and

correct violation(s), and how is this demonstrated and/or documented? See IOM 5.3.6.2.

3. Report the chain of command; include an organizational chart (create if necessary).

4. Obtain a copy of public annual report, if any.

5. List the names and titles of key operating personnel.

5.10.4.3.8 - Firm's Training Program

The firm's training programs are of particular significance where inspectional findings find people may not be adequately trained.

5.10.4.3.9 - Manufacturing/Design Operations

Manufacturing/Design Operations:

1. Report only changes to the firm's general overall operations, including significant changes in equipment, processes, or products since the previous inspection. Include schematics, flow plans, photographs, formulations and diagrams, if useful.

2. List names and sources of new or unusual components or raw materials.

3. Report equipment considered new or unusual unless otherwise directed.

4. Submit pertinent formulas (especially those being manufactured during your inspection) and processing instructions with labeling of suspect products.

For human drug inspection reports:

This section of the EIR should be organized by system covered during the EI as outlined in CPGM 7356.002. In each section, include a brief summary of what you reviewed in order to meet the key system element outlined in the CPGM. You should add more detail for the system elements found to be deficient, or the subject of a FDA 483 observation.

For medical device inspection reports:

1. Describe manufacturing operations by sub system covered in your inspection (Management Controls, Design Controls, Production and Process Controls, Corrective and Preventive Action Controls, Material Controls, Facility and Equipment Controls, and Records/Documents/Change Controls). For ALL Level 2, 3, and "for cause" inspections: for production and process controls - indicate which production processes were covered/reviewed. If a subsystem was not specifically covered during your EI, you do not need to separately describe the general operations of that subsystem.

2. For all inspections covering CAPA - indicate which data sources were available for review and which were actually reviewed; include a brief statement regarding coverage or non-coverage of applicable tracking requirements, MDRs, sterilization, and reports of corrections and removals.

3. If the Design Control system was covered, indicate the design project(s) covered during the inspection. Where design activities occur at a location other than the manufacturing site, list the name, address of the design location and responsibilities of those performing the design activities.

4. If applicable, identify the name and address of the specification developer if different from either the manufacturing site or where design activities occur.

5.10.4.3.10 - Manufacturing Codes

Manufacturing Codes

1. If the manufacturing codes are unchanged, include a statement in the EIR the system is the same as described in reports on file at the District.   Indicate the date of the EIR in which the codes are fully explained.

2. If the manufacturing codes have changed, describe the manufacturing coding system (lot, batch, product, etc.), and a key to interpretation of codes.

3. For medical device inspections reports: where appropriate, include a description of the system used to identify and maintain control of components during the

manufacturing process, as well as, the codes used for traceability (for applicable finished devices).

5.10.4.3.11 - Complaints

Note: These complaints include those reported to the FDA by consumers, health care professionals, industry, etc.; and all complaints received by the firm.

1. Report your review of the firm's complaint file(s).

2. In addition, if returned goods and/or documents for returned goods are examined, describe findings. If not examined, so indicate.

3. Report your follow-up of consumer/trade complaints, Adverse Event Reports, MDR's, MedWatch reports or recalls identified in the district factory jacket for coverage. Correlate consumer/trade complaints, Adverse Event Reports, MDR's, MedWatch reports to specific objectionable conditions observed.

5.10.4.3.12 - Recall Procedures

Describe plans and procedures for removing products from marketing channels if necessary.   If these procedures are in written SOP-type format, you may reference any copies obtained to aid in your explanation.

5.10.4.3.13 - Objectionable Conditions and Management's Response

If any observations were provided to management in writing (FDA 483) at the conclusion of the inspection list each observation and report each observation providing information organized under the two headings Supporting Evidence and Relevance, and Discussion with Management below.

NOTE: Observations of a verbal nature should be reported in sufficient detail under the General Discussion with Management (correlate any Exhibits, samples, etc. to any "verbal" observations).

5.10.4.3.13.1 - Supporting Evidence and Relevance

Sufficiently describe the observation as necessary to relate the facts as you found them.

1. Identify specific pages of exhibits and/or samples (e.g., procedure title, section, paragraph, sentence), labeling text, interstate shipping records which in your judgment document violations so supervisors, compliance officers, and other reviewers can readily evaluate your evidence.

2. Describe verbal statements (verbatim if possible) by firm officials having knowledge, duty, power, and responsibility to detect, prevent, or correct the apparent violation.

3. Identify the responsible party for each apparent violation (i.e., if known.)

4. Identify which team member (if applicable) was responsible for the observation.

5. When appropriate explain how this observation relates to the overall situation; i.e., impact on the product, batches, or lots involved, and any relationship to other products, processes, or other FDA 483 observations.

6. The duration of the problem.

5.10.4.3.13.2 - Discussion with management

Discussion with management:

1. Report management's response to each specific observation, time frames given for corrections and/or corrective action.

2. Report any disagreements with or refusals to correct the observation.

For medical device inspection reports:

1. For each observation based on sampling of records, indicate which Sample Table and level of confidence was used and the actual number of records sampled.

2. If the number sampled is different than the actual number reviewed, so indicate.

5.10.4.3.14 - Refusals

Provide full details of all refusals of/for requested information, statutory information, photography, entry, etc. received during the inspection, including who made the refusal and, if available, why the refusal was given.

5.10.4.3.15 - General Discussion with Management

General Discussion with Management:

1. Report the names and titles of all present, including those present via electronic media (describe).

2. Include the name and title to whom the FDA 483 was issued.

3. Provide additional discussion items not provided in writing at the conclusion of the inspection, such as: questionable labels, labeling and/or labeling practices, commercialization of products covered by IDE or IND, fraudulent health claims, registration/listing deviations, lack of approved PMA, 510(k), NDA, ANDA, etc. These include all verbal observations deemed not to merit inclusion on the FDA 483 (IOM 5.2.3)

4. A description of each warning, recommendation, or suggestion given to the firm, and to whom given.

5. Management's general responses to the inspection and/or to groups of items listed on the report of observations or discussed at the conclusion of the inspection.

5.10.4.3.16 - Additional Information

Report changes as appropriate.

1. Describe contractors used and for what purpose. For Medical Device inspection reports: also include names and addresses of all applicable third party installers or servicing organizations used by the manufacturer. Include their responsibilities.

2. Describe suppliers (major raw material, active ingredient, etc.) used and for what.

3. During inspections, when violative products imported into the U.S., or intended to be imported into the U.S. are encountered, document the product and foreign manufacturer in the EIR. Violative products could be rejected APIs due to non-conformance with the USP, foods without appropriate labeling, etc. Send a copy of the EIR to DIOP (HFC-170). See IOM 5.2.1 and 5.10.2.

4. For initial inspections, verify distribution patterns for the firm's products, raw materials, and components to firms which warehouse or further process products which may be subject to FDA regulations. Districts should incorporate information obtained into their Official Establishment Inventory improvement activities and complete form FDA 457, Product/Establishment Surveillance Report as appropriate. See IOM 8.6.2.

5. Report pertinent facts, which do not fit another section of the EIR. (For firms located in foreign countries, include information relative to lodging and travel; for domestic firms, include information relative to location of firm if difficult to find; etc.).

For human drug inspection reports - PDMA Coverage:

1. Describe what sample loss, theft, or diversion reports were covered during the inspection.

2. Describe the firm's sample audit and security systems, including a review of the firm's SOP's. Significant problems which may contribute to the firm's inability to adequately monitor sample distribution via sales representative, mail or common carrier should be addressed under objectionable conditions.

5.10.4.3.17 - Samples Collected

List and describe samples collected during the inspection.

5.10.4.3.18 - Voluntary Corrections

Voluntary Corrections:

1. Provide a brief description of improvements initiated by the firm in response to a previous inspection, report of observations and/or a warning letter.

2. Report voluntary destructions, recalls, and similar actions since the prior inspection or during this inspection.

3. Report any follow-up to recalls identified during the inspection (may be by referencing Attachment B recall report).

4. Include recalls to specific objectionable conditions observed.

5. Provide the identity of person(s) responsible for the corrections.

6. Report any appropriate voluntary corrections in FACTS CARS.

5.10.4.3.19 - Exhibits Collected

List all exhibits attached. See IOM 5.10.5, Exhibits.

Briefly, describe or title each exhibit and sample number attached. You should include in your description the number of pages for each Exhibit listing.

NOTE: *For complex inspections a cross-reference from the FDA 483 and verbal observations to applicable exhibits and samples can be useful during further review.*

5.10.4.3.20 - Attachments

Attachments as referred to here are any material attached to and referred to in the EIR, which are not evidentiary in nature; such as assignments, Center provided protocols, etc. See IOM 5.3.8.2 for identification of non-evidentiary material attached to the EIR. Documents attached to the EIR may be referred to in the EIR and listed here, such as the FDA 482, FDA 483, copy of the FDA 463a, etc. (in form number order); but such documents/forms may not be numbered, altered from their issued state, bear adhesive identification labels, etc. See the opening sentence of IOM 5.10.5. List and attach copies of associated reports (Recall Attachment B Report, etc.).

5.10.4.3.21 - Signature

All participants will sign the final narrative portion of the EIR. The prescribed format is to type each persons name, title, and district (or other affiliation) below the signature. In some cases immediate signature by all participants is not possible. An example as to how this can be accomplished is to forward an electronic "draft" copy of the EIR for all to read and approve, then followed or accompanied by the original signature sheet. When signed, return to the lead investigator for proper filing and routing. When using this method, a photocopy of the original signature page is made with the lead investigator's signature and temporarily attached to the EIR.

5.10.5 - Exhibits

Exhibits are materials collected from the firm and do not include FDA forms or copies of assignments. Exhibits should contribute to the objective of the assignment and the clarity of the report. They may include flow-plans, schematics, layouts, etc. If the materials collected from the firm are not needed as exhibits, they should be destroyed in accordance with district policy. Submit at least three copies of new or suspect labeling or other material collected as exhibits for labeling purposes. See IOM 4.4.9 for exceptions. These should be mounted in a manner so complete sets are submitted that can be reviewed by individuals in separate offices, i.e., labels 1-10 in each of three sets. You should identify records/exhibits submitted with an EIR using at least the Exhibits' number, firm name, date(s) of the inspection, and your initials. See IOM 5.3.8.2.

5.10.5.1 - Electronic information

Electronic information, databases or summary data from databases may be obtained from firms and evaluated during the course of an EI. This data may form the basis for observations or information included in the EIR. It is preferable to include a printed version and/or a summary of the data as an exhibit. When it is included as an exhibit to the EIR, it should be stored so as to protect the integrity of the data. See IOM 5.3.8.3.2 for procedures for collecting and identifying electronic data. Electronic media should be protected from extreme temperatures and most magnetic fields. Additional precautions may be necessary and you should be guided by your district procedures for storage of electronic data.

5.10.6 - Addendum To EIR

If your EIR requires correcting or clarification after it has been finalized, signed and distributed, you should prepare an addendum, with your supervisor's approval. The addendum should clearly identify itself with the EIR being added to, explain the necessity for the addendum, and clearly define what section(s) and page(s) are being revised. The addendum will be signed by the preparer.

Part II

Exhibits, Appendix, and Samples

IOM Exhibits, Appendix, and Samples

The Investigations Operations Manual (IOM) provides chapter-specific exhibits, appendixes, and samples that demonstrate and support the procedures detailed in the IOM.

Exhibits[17]

Exhibits supporting each chapter of the Investigations Operations Manual are available on the FDA website at http://www.fda.gov/ICECI/Inspections/IOM/ucm127372.htm.

The list of available exhibits for each chapter follows.

Chapter 5 - Exhibits

5-1 Form FDA 482 Notice of Inspection (2 Pgs)

5-2 FDA 482a Demand for Records

5-3 FDA 482b Request for Information

5-4 Modified FDA 482

5-5 Form FDA 483 (2 Pgs)

5-6 Inserting Digital Photos into Turbo EIR (Resize Photo)

5-7 Inserting Digital Photos into Turbo EIR (Insert Photo)

5-8 Inserting Digital Photos into Turbo EIR (Resize Using MS Office Picture Manager)

5-9 Facts Create Assignment Screen (1 Pg)

5-10 FDA 482(C) Notice of Inspection Request for Records

5-11 Food Additives Nomographs

---

[17] Downloadable files are available on the FDA website at http://www.fda.gov/ICECI/Inspections/IOM/ucm127372.htm

5-12 Summary of Registration and Listing*** Human Pharmaceuticals (1 Pg)

5-13 Substantially Equivalent Medical Devices (1 Pg)

5-14 Facts - Profile - Comstat (6 Pgs)

5-15 Compliance Achievement Report

5-16 Facts EI Record (3 Pgs)

Appendixes

Appendixes supporting each chapter of the Investigations Operations Manual are available on the FDA website at http://www.fda.gov/ICECI/Inspections/IOM/ucm127497.htm.

The list of available appendixes for each chapter follows:

Appendix A - Insects

Appendix B - Calendars

Appendix C - Blood Values

Appendix D - Conversion Factors

Appendix E - ORA Map

Sample Schedules

Sample Schedule 1:	Salmonella Sampling Plan
Sample Schedule 2:	Sample Scheduling For Canned And Acidified Foods
Sample Schedule 3:	Pesticide Samples
Sample Schedule 4:	Wheat Carload Sampling
Sample Schedule 5:	Imported Whitefish Sampling Schedule
Sample Schedule 6:	Mycotoxin Sample Sizes
Sample Schedule 7:	Canned Fruit - Fill Of Container - Authentic Pack
Sample Schedule 8:	Imports - Coffee, Dates, and Date Material
Sample Schedule 9:	Sampling Schedule For Color Containing Products Color Additives
Sample Schedule 10:	Drug Sampling Schedules (Does not include Antibiotic Preparations)

Sample Schedule 11: Veterinary Products, Feeds, & By-Products For Animal Feeds

Sample Schedule 12: Medicated Animal Feeds Sampling

Sample Schedule 13: Allergen Sample Schedule

IOM Chapter 5 Exhibits

CONTENT

5-1   Form FDA 482 Notice of Inspection (2 Pgs)

5-2   FDA 482a Demand for Records

5-3   FDA 482b Request for Information

5-4   Modified FDA 482

5-5   Form FDA 483 (2 Pgs)

5-6   Inserting Digital Photos into Turbo EIR (Resize Photo)

5-7   Inserting Digital Photos into Turbo EIR (Insert Photo)

5-8   Inserting Digital Photos into Turbo EIR (Resize Using MS Office Picture Manager)

5-9   Facts Create Assignment Screen (1 Pg)

5-10   FDA 482(C) Notice of Inspection Request for Records

5-11   Food Additives Nomographs

5-12   Summary of Registration and Listing*** Human Pharmaceuticals (1 Pg)

5-13   Substantially Equivalent Medical Devices (1 Pg)

5-14   Facts - Profile - Comstat (6 Pgs)

5-15   Compliance Achievement Report

5-16   Facts EI Record (3 Pgs)

Exhibit 5-1: Form FDA 482 Notice of Inspection (2 Pgs)

DEPARTMENT OF HEALTH AND HUMAN SERVICES FOOD AND DRUG ADMINISTRATION	1. DISTRICT OFFICE ADDRESS & PHONE NO. 1431 Harbor Bay Parkway Alameda, CA 94502-7070 408-225-5332

TO

2. NAME AND TITLE OF INDIVIDUAL Robert K. Thompson, Plant Manager	3. DATE 8-10-08
4. FIRM NAME Garden City Nut Shellers	HOUR 8:30 a.m.
6. NUMBER AND STREET 2704 Sellers Ave	p.m.
7. CITY AND STATE & ZIP CODE San Jose, CA 95131	8. PHONE # & AREA CODE 408-213-4567

Notice of Inspection is hereby given pursuant to Section 704(a)(1) of the Federal Food, Drug, and Cosmetics Act [21 U.S.C. 374(a)][1] and/or Part F or G, Title III of the Public Health Service Act [42 U.S.C. 262-264][2]

9. SIGNATURE (Food and Drug Administration Employee(s))	10. TYPE OR PRINT NAME AND TITLE (FDA Employee(s))
Sidney H. Rogers	Sidney H. Rogers, Investigator

[1] Applicable portions of Section 704 and other Sections of the Federal Food, Drug, and Cosmetic Act [21 U.S.C. 374] are quoted below:

Sec. 704. (a)(1) For purposes of enforcement of this Act, officers or employees duly designated by the Secretary, upon presenting appropriate credentials and a written notice to the owner, operator, or agent in charge, are authorized (A) to enter, at reasonable times, any factory, warehouse, or establishment in which food, drugs, devices, or cosmetics are manufactured, processed, packed, or held, for introduction into interstate commerce or after such introduction, or to enter any vehicle being used to transport or hold such food, drugs, devices, or cosmetics in interstate commerce, and (B) to inspect, at reasonable times and within reasonable limits and in a reasonable manner, such factory, warehouse, establishment, or vehicle and all pertinent equipment, finished and unfinished materials, containers, and labeling therein. In the case of any person (excluding farms and restaurants) who manufactures, processes, packs, transports, distributes, holds, or imports foods, the inspection shall extend to all records and other information described in section 414 when the Secretary has a reasonable belief that an article of food is adulterated and presents a threat of serious adverse health consequences or death to humans or animals, subject to the limitations established in section 414(d). In the case of any factory, warehouse, establishment, or consulting laboratory in which prescription drugs, nonprescription drugs intended for human use, or restricted devices are manufactured, processed, packed, or held, inspection shall extend to all things therein (including records, files, papers, processes, controls, and facilities) bearing on whether prescription drugs, nonprescription drugs intended for human use or, restricted devices which are adulterated or misbranded within the meaning of this Act, or which may not be manufactured, introduced into interstate commerce, or sold, or offered for sale by reason of any provision of this Act, have been or are being manufactured, processed, packed, transported, or held in any such place, or otherwise bearing on violation of this Act. No inspection authorized by the preceding sentence or by paragraph (3) shall extend to financial data, sales data other than shipment data, pricing data, personnel data (other than data as to qualifications of technical and professional personnel performing functions subject to this Act), and research data (other than data relating to new drugs, antibiotic drugs and devices and, subject to reporting and inspection under regulations lawfully issued pursuant to section 505(i) or (k), section 519, or 520(g), and data relating to other drugs or devices which in the case of a new drug would be subject to reporting or inspection under lawful regulations issued pursuant to section 505(j)). A separate notice shall be given for each such inspection, but a notice shall not be required for each entry made during the period covered by the inspection. Each such inspection shall be commenced and completed with reasonable promptness.

Sec. 704. (a)(2) The provisions of the third sentence of paragraph (1) shall not apply to (A) pharmacies which maintain establishments in conformance with any applicable local laws regulating the practice of pharmacy and medicine and which are regularly engaged in dispensing prescription drugs or devices, upon prescriptions of practitioners licensed to administer such drugs or devices to patients under the care of such practitioners in the course of their professional practice, and which do not, either through a subsidiary or otherwise, manufacture,

prepare, propagate, compound, or process drugs or devices for sale other than in the regular course of their business of dispensing or selling drugs or devices at retail; (B) practitioners licensed by law to prescribe or administer drugs, or prescribe or use devices, as the case may be, and who manufacture, prepare, propagate, compound, or process drugs, or manufacture or process devices solely for use in the course of their professional practice, (C) persons who manufacture, prepare, propagate, compound, or process drugs, or manufacture or process devices solely for use in research, teaching, or chemical analysis and not for sale; (D) such other classes of persons as the Secretary may by regulation exempt from the application of this section upon a finding that inspection as applied to such classes of persons in accordance with this section is not necessary for the protection of the public health.

Sec. 704. (a)(3) An officer or employee making an inspection under paragraph (1) for purposes of enforcing the requirements of section 412 applicable to infant formulas shall be permitted, at all reasonable times, to have access to and to copy and verify any records (A) bearing on whether the infant formula manufactured or held in the facility inspected meets the requirements of section 412, or (B) required to be maintained under section 412.

Sec. 704. (b) Upon completion of any such inspection of a factory, warehouse, consulting laboratory, or other establishment, and prior to leaving the premises, the officer or employee making the inspection shall give to the owner, operator, or agent in charge a report in writing setting forth any conditions or practices observed by him which, in his judgment, indicate that any food, drug, device, or cosmetic in such establishment (1) consists in whole or in part of any filthy, putrid, or decomposed substance, or (2) has been prepared, packed, or held under insanitary conditions whereby it may have become contaminated with filth, or whereby it may have been rendered injurious to health. A copy of such report shall be sent promptly to the Secretary.

Sec. 704. (c) If the officer or employee making any such inspection of a factory, warehouse, or other establishment has obtained any sample in the course of the inspection, upon completion of the inspection and prior to leaving the premises he shall give to the owner, operator, or agent in charge a receipt describing the samples obtained.

Sec. 704. (d) Whenever in the course of any such inspection of a factory or other establishment where food is manufactured, processed, or packed, the officer or employee making the inspection obtains a sample of any such food, and an analysis is made of such sample for the purpose of ascertaining whether such food consists in whole or in part of any filthy, putrid, or decomposed substance, or is otherwise unfit for food, a copy of the results of such analysis shall be furnished promptly to the owner, operator, or agent in charge.

Sec. 704(e) Every person required under section 519 or 520(g) to maintain records and every person who is in charge or custody of such records shall, upon request of an officer or employee designated by the Secretary, permit such officer or employee at all reasonable times to have access to and to copy and verify, such records

Section 704 (f)(1) An accredited person described in paragraph (3) shall maintain records documenting the training qualifications of the person

FORM FDA 482 (4/08)   PREVIOUS EDITION IS OBSOLETE   (Continued on Reverse)   NOTICE OF INSPECTION

PSC Media Arts (301) 443-1090 EF

Exhibit 5-1: Form FDA 482 Notice of Inspection (2 Pgs)

and the employees of the person, the procedures used by the person for handling confidential information, the compensation arrangements made by the person, and the procedures used by the person to identify and avoid conflicts of interest. Upon the request of an officer or employee designated by the Secretary, the person shall permit the officer or employee, at all reasonable times, to have access to, to copy, and to verify, the records.

Section 512 (l)(1) In the case of any new animal drug for which an approval of an application filed pursuant to subsection (b) is in effect, the applicant shall establish and maintain such records, and make such reports to the Secretary, of data relating to experience, including experience with uses authorized under subsection (a)(4)(A), and other data or information, received or otherwise obtained by such applicant with respect to such drug, or with respect to animal feeds bearing or containing such drug, as the Secretary may by general regulation, or by order with respect to such application, prescribe on the basis of a finding that such records and reports are necessary in order to enable the Secretary to determine, or facilitate a determination, whether there is or may be ground for invoking subsection (e) or subsection (m)(4) of this section. Such regulation or order shall provide, where the Secretary deems it to be appropriate, for the examination, upon request, by the persons to whom such regulation or order is applicable, of similar information received or otherwise obtained by the Secretary.

(2) Every person required under this subsection to maintain records, and every person in charge or custody thereof, shall, upon request of an officer or employee designated by the Secretary, permit such officer or employee at all reasonable times to have access to and copy and verify such records.

[2] Applicable sections of Parts F and G of Title III Public Health Service Act [42 U.S.C. 262-264] are quoted below:

Part F - Licensing - Biological Products and Clinical Laboratories and * * * * *

Sec. 351(c) "Any officer, agent, or employee of the Department of Health and Human Services, authorized by the Secretary for the purpose, may during all reasonable hours enter and inspect any establishment for the propagation or manufacture and preparation of any virus, serum, toxin, antitoxin, vaccine, blood, blood component or derivative, allergenic product, or other product aforesaid for sale, barter, or exchange in the District of Columbia, or to be sent, carried, or brought from any State or possession into any other State or possession or into any foreign country, or from any foreign country into any State or possession."
Part F - * * * * * "Control of Radiation.
Sec. 360 A (a) "If the Secretary finds for good cause that the methods, tests, or programs related to electronic product radiation safety in a particular factory, warehouse, or establishment in which electronic products are manufactured or held, may not be adequate or reliable, officers or employees duly designated by the Secretary, upon presenting appropriate credentials and a written notice to the owner, operator, or agent in charge, are thereafter authorized (1) to enter, at reasonable times any area in such factory, warehouse, or establishment in which the manufacturer's tests (or testing programs) required by section 358(h) are carried out, and (2) to inspect, at reasonable times and within reasonable limits and in a reasonable manner, the facilities and procedures within such area which are related to electronic product radiation safety. Each such inspection shall be commenced and completed with reasonable promptness. In addition to other grounds upon which good cause may be found for purposes of this subsection, good cause will be considered to exist in any case where the manufacturer has introduced into commerce any electronic product which does not comply with an applicable standard prescribed under this subpart and with respect to which no exemption from the notification requirements has been granted by the Secretary under section 359(a)(2) or 359(e)."

(b) "Every manufacturer of electronic products shall establish and maintain such records (including testing records), make such reports, and provide such information, as the Secretary may reasonably require to enable him to determine whether such manufacturer has acted or is acting in compliance with this subpart and standards prescribed pursuant to this subpart and shall, upon request of an officer or employee duly designated by the Secretary, permit such officer or employee to inspect appropriate books, papers, records, and documents relevant to determining whether such manufacturer has acted or is acting in compliance with standards prescribed pursuant to section 359(a)."

* * * * * *

(f) "The Secretary may by regulation (1) require dealers and distributors of electronic products, to which there are applicable standards prescribed under this subpart and the retail prices of which is not less than $50, to furnish manufacturers of such products such information as may be necessary to identify and locate, for purposes of section 359, the first purchasers of such products for purposes other than resale, and (2) require manufacturers to preserve such information. Any regulation establishing a requirement pursuant to clause (1) of the preceding sentence shall (A) authorize such dealers and distributors to elect, in lieu of immediately furnishing such information to the manufacturer to hold and preserve such information until advised by the manufacturer or Secretary that such information is needed by the manufacturer for purposes of section 359, and (B) provide that the dealer or distributor shall, upon making such election, give prompt notice of such election (together with information identifying the notifier and the product) to the manufacturer and shall, when advised by the manufacturer or Secretary, of the need therefore for the purposes of Section 359, immediately furnish the manufacturer with the required information. If a dealer or distributor discontinues the dealing in or distribution of electronic products, he shall turn the information over to the manufacturer. Any manufacturer receiving information pursuant to this subsection concerning first purchasers of products for purposes other than resale shall treat it as confidential and may use it only if necessary for the purpose of notifying persons pursuant to section 359(a)."

* * * * * *

Sec. 360 B (a) It shall be unlawful-
(1) * * *
(2) * * *
(3) "for any person to fail or to refuse to establish or maintain records required by this subpart or to permit access by the Secretary or any of his duly authorized representatives to, or the copying of, such records, or to permit entry or inspection, as required by or pursuant to section 360A."

* * * * * *

Part G - Quarantine and Inspection

Sec. 361(a) "The Surgeon General, with the approval of the Secretary, is authorized to make and enforce such regulations as in his judgment are necessary to prevent the introduction, transmission, or spread of communicable diseases from foreign countries into the States or possessions, or from one State or possession into any other State or possession. For purposes of carrying out and enforcing such regulations, the Surgeon General may provide for such inspection, fumigation, disinfection, sanitation, pest extermination, destruction of animals or articles found to be so infected or contaminated as to be sources of dangerous infection to human beings, and other measures, as in his judgment may be necessary."

(Reverse of Form FDA 482)

FORM FDA 482 (4/08)

Exhibit 5-2: FDA 482a Demand for Records

DEPARTMENT OF HEALTH AND HUMAN SERVICES FOOD AND DRUG ADMINISTRATION	1. DISTRICT ADDRESS AND PHONE NO. 6751 Steger Dr. Cincinnati, OH 45237 (513) 679-2700		

TO	2. NAME AND TITLE OF INDIVIDUAL Michael Campbell, President		3. DATE 8/20/2007	
	4. FIRM NAME ABC Soup Company		5. HOURS 8:30	
	6. NUMBER AND STREET 3114 Mapleleaf Ave		A.M. XXX	P.M.
	7. CITY AND STATE Cincinnati, OH		8. ZIP CODE 45213	

Written demand for examination and/or copying of the records required by 21 CFR 113.100, 21 CFR 114 and 21 CFR 500.23 is hereby given, pursuant to 21 CFR 108.25(g), 21 CFR 108.35(h) and 21 CFR 500 for the records described below in order to verify the pH, adequacy of processing, the integrity of container closures, and the coding of the products processed by your firm.

9. RECORDS NECESSARY

All thermal process and production records mandated by 21 CFR 113 and 114 for the manufacture of Mighty Good Vegetable Soup in 303 cans from May 1, 2004 to the present. This includes pH records, calibration record, formulation, batch records, etc.

10. SIGNATURE *(Food and Drug Administration Employee(s))* *Sydney H. Rogers*	11. TITLE FDA EMPLOYEE Investigator

FORM FDA 482a (10/03)　　　　PREVIOUS EDITION IS OBSOLETE　　　　DEMAND FOR RECORDS

Exhibit 5-3: FDA 482b Request for Information

Exhibit 5-3: FDA 482b Request for Information

DEPARTMENT OF HEALTH AND HUMAN SERVICES FOOD AND DRUG ADMINISTRATION	1. DISTRICT ADDRESS AND PHONE NO. 6751 Steger Dr. Cincinnati, OH 45237 (513) 679-2700	

TO	2. NAME AND TITLE OF INDIVIDUAL Michael Campbell, President		3. DATE 8/20/2007	
	4. FIRM NAME ABC Company		5. HOURS 8:00	
	6. NUMBER AND STREET 3114 Mapleleaf Ave.		A.M. xxx	P.M.
	7. CITY AND STATE Cincinnati, OH		8. ZIP CODE 45213	

Written request is hereby given pursuant to 21 CFR 108.25(c)(3)(ii), 21 CFR 108.35(c)(3)(ii) and 21 CFR 500.23 for the information described below, concerning processes and procedures which is deemed necessary by the Food and Drug Administration to determine the adequacy of the processes for products processed by your firm.

9. RECORDS NECESSARY

All written supporting documentation from a process authority or other source which specify the scheduled process and critical factors for the processing of Mighty Good Vegetable Soup in 303 cans to include pH records, calibration records, formulation and batch records, etc

10. SIGNATURE (Food and Drug Administration Employee(s)) Sydney H. Rogers	11. TITLE FDA EMPLOYEE Investigator

FORM FDA 482b (10/03)   PREVIOUS EDITION IS OBSOLETE.   **REQUEST FOR INFORMATION**

Exhibit 5-4: Modified FDA 482

DEPARTMENT OF HEALTH AND HUMAN SERVICES FOOD AND DRUG ADMINISTRATION	1. DISTRICT OFFICE ADDRESS & PHONE NO. 22201 23rd Drive SE Seattle, WA 98021-4421 503-225-5332

TO	2. NAME AND TITLE OF INDIVIDUAL Howard M. Allgreen, Pharmacist-Owner	3. DATE 8-10-08
	4. FIRM NAME Darlings Drug Store	5. HOUR a.m. 2:00 p.m.
	6. NUMBER AND STREET 312 Main Street	
	7. CITY AND STATE & ZIP CODE Medford, OR 97501	8. PHONE # & AREA CODE 503-765-4321

Notice of Inspection is hereby given to collect samples only pursuant to Section 704(a)(1) of the Federal Food, Drug, and Cosmetics Act [21 U.S.C. 374(a)][1] and/or Part F or G, Title III of the Public Health Service Act [42 U.S.C. 262-264][2]

9. SIGNATURE *(Food and Drug Administration Employee(s))* *Sidney H. Rogers*	10. TYPE OR PRINT NAME AND TITLE *(FDA Employee(s))* Sidney H. Rogers, Investigator

[1] Applicable portions of Section 704 and other Sections of the Federal Food, Drug, and Cosmetic Act [21 U.S.C. 374] are quoted below:

Sec. 704 (a)(1) For purposes of enforcement of this Act, officers or employees duly designated by the Secretary, upon presenting appropriate credentials and a written notice to the owner, operator, or agent in charge, are authorized (A) to enter, at reasonable times, any factory, warehouse, or establishment in which food, drugs, devices, or cosmetics are manufactured, processed, packed, or held, for introduction into interstate commerce or after such introduction, or to enter any vehicle being used to transport or hold such food, drugs, devices, or cosmetics in interstate commerce; and (B) to inspect, at reasonable times and within reasonable limits and in a reasonable manner, such factory, warehouse, establishment, or vehicle and all pertinent equipment, finished and unfinished materials, containers, and labeling therein. In the case of any person (excluding farms and restaurants) who manufactures, processes, packs, transports, distributes, holds, or imports foods, the inspection shall extend to all records and other information described in section 414 when the Secretary has a reasonable belief that an article of food is adulterated and presents a threat of serious adverse health consequences or death to humans or animals, subject to the limitations established in section 414(d). In the case of any factory, warehouse, establishment, or consulting laboratory in which prescription drugs, nonprescription drugs intended for human use, or restricted devices are manufactured, processed, packed, or held, inspection shall extend to all things therein (including records, files, papers, processes, controls, and facilities) bearing on whether prescription drugs, nonprescription drugs intended for human use or restricted devices which are adulterated or misbranded within the meaning of this Act, or which may not be manufactured, introduced into interstate commerce, or sold, or offered for sale by reason of any provision of this Act, have been or are being manufactured, processed, packed, transported, or held in any such place, or otherwise bearing on violation of this Act. No inspection authorized by the preceding sentence or by paragraph (3) shall extend to financial data, sales data other than shipment data, pricing data, personnel data (other than data as to qualifications of technical and professional personnel performing functions subject to this Act), and research data (other than data relating to new drugs, antibiotic drugs and devices and, subject to reporting and inspection under regulations lawfully issued pursuant to section 505(i) or (k), section 519, or 520(g), and data relating to other drugs or devices which in the case of a new drug would be subject to reporting or inspection under lawful regulations issued pursuant to section 505(j)). A separate notice shall be given for each such inspection, but a notice shall not be required for each entry made during the period covered by the inspection. Each such inspection shall be commenced and completed with reasonable promptness.

Sec. 704 (a)(2) The provisions of the third sentence of paragraph (1) shall not apply to (A) pharmacies which maintain establishments in conformance with any applicable local laws regulating the practice of pharmacy and medicine and which are regularly engaged in dispensing prescription drugs or devices, upon prescriptions of practitioners licensed to administer such drugs or devices to patients under the care of such practitioners in the course of their professional practice, and which do not, either through a subsidiary or otherwise, manufacture,

prepare, propagate, compound, or process drugs or devices for sale other than in the regular course of their business of dispensing or selling drugs or devices at retail; (B) practitioners licensed by law to prescribe or administer drugs, or prescribe or use devices, as the case may be, and who manufacture, prepare, propagate, compound, or process drugs, or manufacture or process devices solely for use in the course of their professional practice; (C) persons who manufacture, prepare, propagate, compound, or process drugs, or manufacture or process devices solely for use in research, teaching, or chemical analysis and not for sale; (D) such other classes of persons as the Secretary may by regulation exempt from the application of this section upon a finding that inspection as applied to such classes of persons in accordance with this section is not necessary for the protection of the public health.

Sec. 704. (a)(3) An officer or employee making an inspection under paragraph (1) for purposes of enforcing the requirements of section 412 applicable to infant formulas shall be permitted, at all reasonable times, to have access to and to copy and verify any records (A) bearing on whether the infant formula manufactured or held in the facility inspected meets the requirements of section 412, or (B) required to be maintained under section 412.

Sec. 704. (b) Upon completion of any such inspection of a factory, warehouse, consulting laboratory, or other establishment, and prior to leaving the premises, the officer or employee making the inspection shall give to the owner, operator, or agent in charge a report in writing setting forth any conditions or practices observed by him which, in his judgment, indicate that any food, drug, device, or cosmetic in such establishment (1) consists in whole or in part of any filthy, putrid, or decomposed substance, or (2) has been prepared, packed, or held under insanitary conditions whereby it may have become contaminated with filth, or whereby it may have been rendered injurious to health. A copy of such report shall be sent promptly to the Secretary.

Sec. 704. (c) If the officer or employee making any such inspection of a factory, warehouse, or other establishment has obtained any sample in the course of the inspection, upon completion of the inspection and prior to leaving the premises he shall give to the owner, operator, or agent in charge a receipt describing the samples obtained.

Sec. 704. (d) Whenever in the course of any such inspection of a factory or other establishment where food is manufactured, processed, or packed, the officer or employee making the inspection obtains a sample of any such food, and an analysis is made of such sample for the purpose of ascertaining whether such food consists in whole or in part of any filthy, putrid, or decomposed substance, or is otherwise unfit for food, a copy of the results of such analysis shall be furnished promptly to the owner, operator, or agent in charge.

Sec. 704(e) Every person required under section 519 or 520(g) to maintain records and every person who is in charge or custody of such records shall, upon request of an officer or employee designated by the Secretary, permit such officer or employee at all reasonable times to have access to and to copy and verify, such records.

Section 704 (f)(1) An accredited person described in paragraph (3) shall maintain records documenting the training qualifications of the person

FORM FDA 482 (4/08)   PREVIOUS EDITION IS OBSOLETE   (Continued on Reverse)   NOTICE OF INSPECTION

PSC Media Arts (301) 443-1090   FF

Exhibit 5-5: Form FDA 483 (2 Pgs)

Exhibit 5-5: Form FDA 483 (2 Pgs)

DEPARTMENT OF HEALTH AND HUMAN SERVICES
FOOD AND DRUG ADMINISTRATION

DISTRICT OFFICE ADDRESS AND PHONE NUMBER	DATE(S) OF INSPECTION
Minneapolis District	10/5-7/2008
250 Marquette Ave. South, Suite 600	**FEI NUMBER**
Minneapolis, MN 55401	0000112233
Industry information: www.fda.gov/oc/industry	

NAME AND TITLE OF INDIVIDUAL TO WHOM REPORT IS ISSUED

TO: William S. Gundstrom, Vice President, Production

FIRM NAME	STREET ADDRESS
Topline Pharmaceuticals "T.L.P."	2136 Elbe Place

CITY, STATE AND ZIP CODE	TYPE OF ESTABLISHMENT INSPECTED
Jackson, MN 55326	Tablet Repacker

THIS DOCUMENT LISTS OBSERVATIONS MADE BY THE FDA REPRESENTATIVE(S) DURING THE INSPECTION OF YOUR FACILITY. THEY ARE INSPECTIONAL OBSERVATIONS, AND DO NOT REPRESENT A FINAL AGENCY DETERMINATION REGARDING YOUR COMPLIANCE. IF YOU HAVE AN OBJECTION REGARDING AN OBSERVATION, OR HAVE IMPLEMENTED, OR PLAN TO IMPLEMENT CORRECTIVE ACTION IN RESPONSE TO AN OBSERVATION, YOU MAY DISCUSS THE OBJECTION OR ACTION WITH THE FDA REPRESENTATIVE(S) DURING THE INSPECTION OR SUBMIT THIS INFORMATION TO FDA AT THE ADDRESS ABOVE. IF YOU HAVE ANY QUESTIONS, PLEASE CONTACT FDA AT THE PHONE NUMBER AND ADDRESS ABOVE.

DURING AN INSPECTION OF YOUR FIRM (I (WE) OBSERVED:

List your observations in a logical manner

See IOM 5.2.3, 5.2.3.1, 5.2.3.2, and 5.2.3.3

SEE REVERSE OF THIS PAGE	EMPLOYEE(S) SIGNATURE	EMPLOYEE(S) NAME AND TITLE (Print or Type)	DATE ISSUED
	Sidney H. Rogers	Sidney H. Rogers, Investigator	10/7/2008

FORM FDA 483 (9/08)   PREVIOUS EDITION OBSOLETE   **INSPECTIONAL OBSERVATIONS**   PAGE 1 of 1 PAGES

The observations of objectionable conditions and practices listed on the front of this form are reported:

1. Pursuant to Section 704(b) of the Federal Food, Drug and Cosmetic Act, or

2. To assist firms inspected in complying with the Acts and regulations enforced by the Food and Drug Administration.

Section 704(b) of the Federal Food, Drug, and Cosmetic Act (21 USC 374(b)) provides:

"Upon completion of any such inspection of a factory, warehouse, consulting laboratory, or other establishment, and prior to leaving the premises, the officer or employee making the inspection shall give to the owner, operator, or agent in charge a report in writing setting forth any conditions or practices observed by him which, in his judgement, indicate that any food, drug, device, or cosmetic in such establishment (1) consists in whole or in part of any filthy, putrid, or decomposed substance, or (2) has been prepared, packed, or held under insanitary conditions whereby it may have become contaminated with filth, or whereby it may have been rendered injurious to health. A copy of such report shall be sent promptly to the Secretary."

Exhibit 5-6: Inserting Digital Photos into Turbo EIR (Resize Photo)

This document is available on the FDA website at
http://www.fda.gov/ICECI/Inspections/IOM/ucm127372.htm.

Exhibit 5-7: Inserting Digital Photos into Turbo EIR (Insert Photo)

This document is available on the FDA website at
http://www.fda.gov/ICECI/Inspections/IOM/ucm127372.htm.

Exhibit 5-8: Inserting Digital Photos into Turbo EIR (Resize Using MS Office Picture Manager)

This document is available on the FDA website at
http://www.fda.gov/ICECI/Inspections/IOM/ucm127372.htm.

Exhibit 5-9: Facts Create Assignment Screen (1 Pg)

Exhibit 5-9: Facts Create Assignment Screen (1 Pg)

FACTS Version 4.9.01 - [Create Assignment]

Action Edit Options Related Info Reports Navigate Tracing FACTS Help Window Help

Assignment

Requesting Organization: MIN-IB-SPV   Requester Completion Date: 11/21/2005   Priority: High   ORA Reqd: N/A   Compliance Number: 145 - 0

Subject: Pharma-Mix Digitalis   ORA Cncrnc Num: Survey Num:

Remarks: Request for samples as follow-up to violative EI

Background Information: Firm's manufacturing practices may lead to low potency

Reporting Method: Notify SI DoMore by phone when accomplish

POC Name: James W. DoMore

Operations

Op Code: 31 Sample Collection   Rqstr Prty: High

Subject: Pharma-Mix Digitalis   Target Date: 11/21/2005

Requester Remarks: Collect 12/100 tab bottles of lot DC-01234 as follow-up to vi

Estmtd Hours: 4   Estmtd Smpl Cost: $150.00   Reimbursable:

Organizations

Accomp Org / Num Of Ops / Perf Org (Adhoc Work)

NMU-DO   1

Add   Next Op   Prev Op   Delete   Delete

PACS & Products

PAC: 56002   Description: DRUG PROCESS INSPEC   Add   Delete

Product Code: 63 F C E 04   Description: Digitalis (Cardiotonic), Human - Rx2   Add   Delete

Firms & Cross References

FEI: 8   Name: 3000900992 Drug Distributors, I   Add   Delete

Code   Value   Add   Delete

Lab References

Sample Number: Delete

PAF:

LID:

FACTS Editor   ✕

Collect 12/100 tab bottles of lot DC-01234
as follow-up to violative EI of Pharma-Mix,
Minneapolis, MN (FEI 3000901012), conducted
on 9/31-10/05/2005. 30 cases were shipped to
Drug Distributers Inc., 3910 Riverside
St.,Newark, NJ on 10/03/05 via Cross Country
Express, Kansas City, MO. Invoice # 8328
10/05/05, B/L A-3026, 10-3-05.|

Search   OK   Cancel

Exhibit 5-10: FDA 482(C) Notice of Inspection Request for Records

DEPARTMENT OF HEALTH AND HUMAN SERVICES FOOD AND DRUG ADMINISTRATION	1. DISTRICT OFFICE ADDRESS & PHONE NO. 212 3rd Avenue South Minneapolis, MN 55401 612-334-4100

TO

2. NAME AND TITLE OF INDIVIDUAL	3. DATE
Howard P. Gunderson, Plant Manager	1/4/06

4. FIRM NAME	5. HOUR
Happy Farms Cheese	7:30 a.m.

6. NUMBER AND STREET	
123 County Road 23	p.m.

7. CITY AND STATE & ZIP CODE	8. PHONE # & AREA CODE
Eau Claire, WI 54703	715-223-4567

Notice of Inspection is hereby given pursuant to Section 704(a)(1) of the Federal Food, Drug, and Cosmetic Act [21 U.S.C. 374(a)(1)][1]. Written request is hereby given to access and/or copy the records described below, pursuant to the Federal Food, Drug and Cosmetic Act, Section 414(a) [21 U.S.C. 350c][2] and Title 21 Code of Federal Regulations, Section 1.361[3].

9. SIGNATURE *(Food and Drug Administration Employee(s))*	10. TYPE OR PRINT NAME AND TITLE *(FDA Employee(s))*
Sidney H. Rogers	Sidney H. Rogers, Investigator

Applicable portions of Sections 704 and 414 of the Federal Food, Drug and Cosmetic Act (21 U.S.C. 374 and 350c) and Title 21 of the Code of Federal Regulations, are quoted below:

[1]Sec. 704 (a)(1) For purposes of enforcement of this Act, officers or employees duly designated by the Secretary, upon presenting appropriate credentials and a written notice to the owner, operator, or agent in charge, are authorized (A) to enter, at reasonable times, any factory, warehouse, or establishment in which food, drugs, devices, or cosmetics are manufactured, processed, packed, or held, for introduction into interstate commerce or after such introduction, or to enter any vehicle being used to transport or hold such food, drugs, devices, or cosmetics in interstate commerce, and (B) to inspect, at reasonable times and within reasonable limits and in a reasonable manner, such factory, warehouse, establishment, or vehicle and all pertinent equipment, finished and unfinished materials, containers and labeling therein. In the case of any person (excluding farms and restaurants) who manufactures, processes, packs, transports, distributes, holds, or imports foods, the inspection shall extend to all records and other information described in section 414 when the secretary has a reasonable belief that an article of food is adulterated and presents a threat of serious adverse health consequences or death to humans or animals, subject to the limitations established in section 414(d). In the case of any factory, warehouse, establishment, or consulting laboratory in which prescription drugs or restricted devices are manufactured, processed, packed, or held, the inspection shall extend to all things therein (including records, files, papers, processes, controls, and facilities) bearing on whether prescription drugs or restricted devices which are adulterated or misbranded within the meaning of this Chapter, or which may not be manufactured, introduced into interstate commerce, or sold, or offered for sale by reason of any provision of this Chapter, have been or are being manufactured, processed, packed, transported, or held in any such place, or otherwise bearing on violation of this Chapter. No inspection authorized by the preceding sentence or by paragraph (3) shall extend to financial data, sales data other than shipment data, pricing data, personnel data (other than data as to qualifications of technical and professional personnel performing functions subject to this Act), and research data (other than data, relating to new drugs, antibiotic drugs and devices and, subject to reporting and inspection under regulations lawfully issued pursuant to section 505(i) or (k), section 507(d) or (g), section 519, or 520(g), and data relating to other drugs or devices which in the case of a new drug would be subject to reporting or inspection under lawful regulations issued pursuant to section 505(k) of this title. A separate notice shall be given for each such inspection, but a notice shall not be required for each entry made during the period covered by the inspection. Each such inspection shall be commenced and completed with reasonable promptness.

[2]Sec. 414(a) Records Inspection. If the Secretary has a reasonable belief that an article of food is adulterated and presents a threat of serious adverse health consequences or death to humans or animals, each person (excluding farms and restaurants) who manufactures, processes, packs, distributes, receives, holds, or imports such article shall, at the request of an officer or employee duly designated by the Secretary, permit such officer or employee, upon presentation of appropriate credentials and a written notice to such person, at reasonable times and within reasonable limits and in a reasonable manner, to have access to and copy all records relating to such article that are needed to assist the Secretary in determining whether the food is adulterated and presents a threat of serious adverse health consequences or death to humans or animals. The requirement under the preceding sentence applies to all records relating to the manufacture, processing, packing, distribution, receipt, holding, or importation of such article maintained by or on behalf of such person in any format (including paper and electronic formats) and at any location).

[3]21 CFR 1.361 What are the record availability requirements? When FDA has a reasonable belief that an article of food is adulterated and presents a threat of serious adverse health consequences or death to humans or animals, any records and other information accessible to FDA under section 414 or 704(a) of the act (21 U.S.C. 350c and 374(a)) must be made readily available for inspection and photocopying or other means of reproduction. Such records and other information must be made available as soon as possible, not to exceed 24 hours from the time of receipt of the official request, from an officer or employee duly designated by the Secretary of Health and Human Services who presents appropriate credentials and a written notice.

FORM FDA 482c (12/05)   **NOTICE OF INSPECTION - REQUEST FOR RECORDS**

Exhibit 5-11: Food Additives Nomographs

Exhibit 5-11: Food Additives Nomographs

FOOD ADDITIVES NOMOGRAPH I

1. Additive and batch weight known. Apply a straight edge to appropriate points on outside columns. Read ppm and/or percent additive where straight edge intersects central column.

2. Tolerance and batch weight known. Apply a straight edge to appropriate points on central and right-hand columns. Read the amount of additive in lbs. or gals. where straight edge intersects the left-hand column.

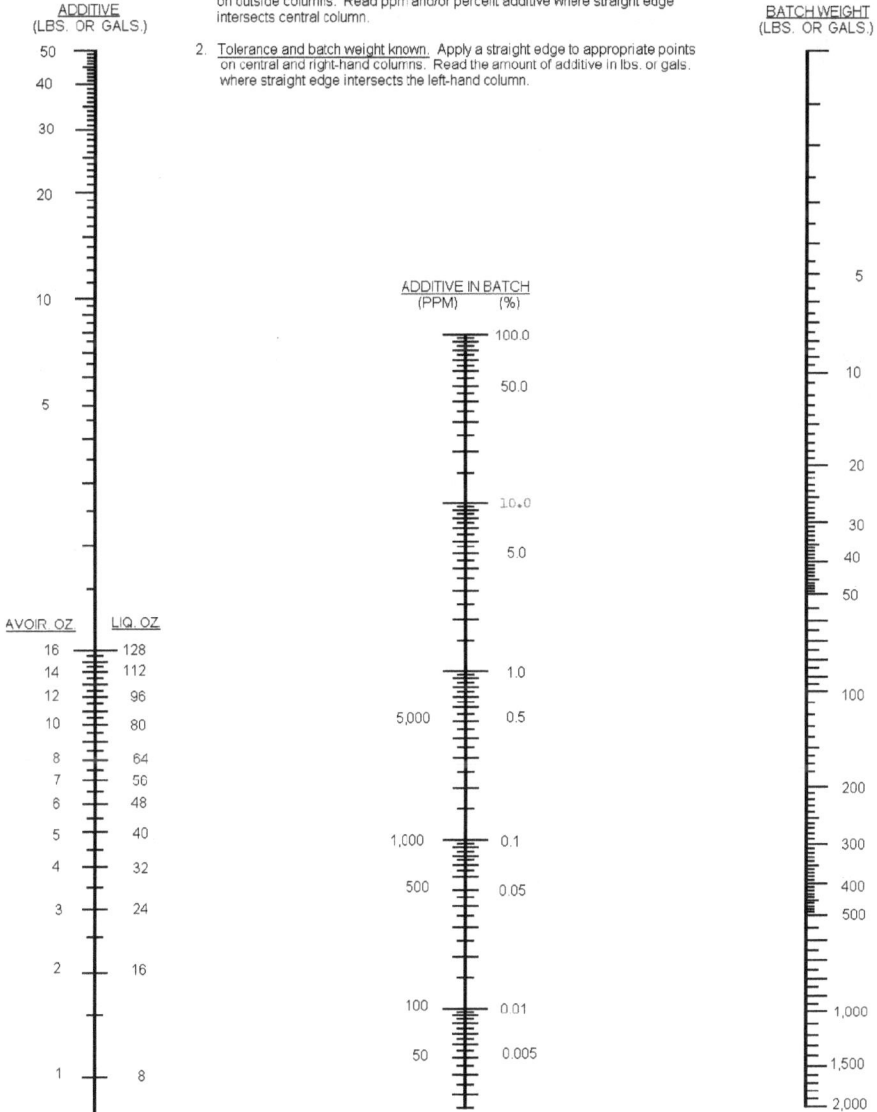

ADDITIVE (LBS. OR GALS.)

BATCH WEIGHT (LBS. OR GALS.)

ADDITIVE IN BATCH (PPM) (%)

AVOIR. OZ | LIQ. OZ

For more precise determination of additives in the 1-500 ppm range, use Nomograph II

FOOD ADDITIVES NOMOGRAPH II

ADDITIVE
(LBS. OR GALS.)

BATCH WEIGHT
(LBS. OR GALS.)

1. Additive and batch weight known. Apply a straight edge to appropriate points on outside columns. Read ppm and/or percent additive where straight edge intersects central column.

2. Tolerance and batch weight known. Apply a straight edge to appropriate points on central and right-hand columns. Read the amount of additive in lbs. or gals. where straight edge intersects the left-hand column.

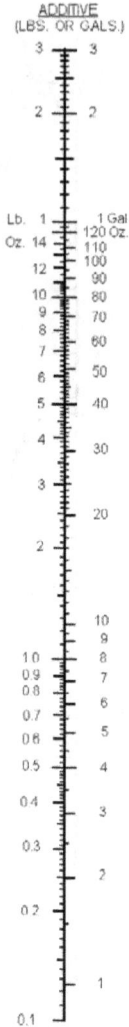

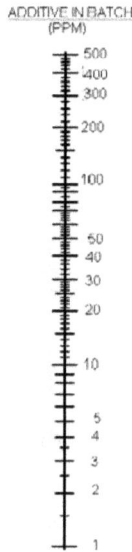

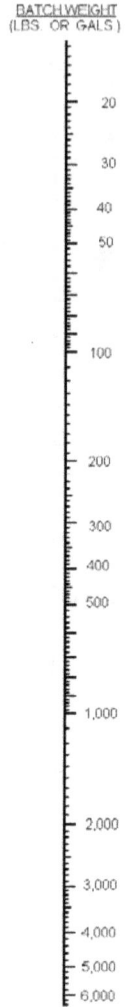

ADDITIVE IN BATCH
(PPM)

Exhibit 5-12: Summary of Registration and Listing*** Human Pharmaceuticals (1 Pg)

SUMMARY OF REGISTRATION AND LISTING REQUIREMENTS FOR THE MANUFACTURE OR DISTRIBUTION OF HUMAN PHARMACEUTICALS			
TYPE OF FIRM	REGISTRATION STATUS	LISTING STATUS	FACTS CODE
Manufacturer [including homeopathic & controlled drugs]	yes	yes	M
Contract Manufacturer	yes	yes	M
Own Label Distributor	no	yes	L
Wholesale Distributor	no	no	W-*
Own Label Repacker	yes	yes	R
Own Label Relabeler [including recirculizer]	yes	yes	Y
Contract Relabeler	yes	no	Y
Contract Testing Laboratory [dosage forms & active ingredient release]	yes	no	C
Contract Testing Lab [doing non-release tests]	no	no	C
Contract Sub-Manufacturer	yes	no	M
IND Manufacturer [Clinical Drugs]	no	no	M
NDA and ANDA Manufacturer	yes	yes	M
Sponsor/Monitors/Clinical Investigator	no	no	4, 5, 6, 7
Contract Sterilizer	yes	no	0
Fulfillment Packager [adding substantive labeling]	yes	no	Y
Mail Order House [adding insubstantial labeling]	no	no	D
Printing House	no	no	None
Medical Gas Transfiller	yes	yes	MG
First Aid/Rescue Squad [transfilling for own use]	no	no	MG
Medical Gas Transfiller [operating out of a van]	yes	yes	MG
Contract Assembler	yes	no	M

Active Drug Substance Manufacturer	yes	yes	M
Excipient Drug Manufacturer	no	no	M
Manufacturer of Research Drugs	no	no	M
Drug Importer	no	no	A
Foreign Drug Manufacturer	yes	yes	M
Methadone Clinic	no	no	T
Retail Pharmacy	no	no	D
Manufacturing Pharmacy	yes	yes	M
Regional Admixture Pharmacy	yes	no	M
Salvage Operation	yes	no	X
Biopharmaceutical Clinical Facility	no	no	2

Exhibit 5-13: Substantially Equivalent Medical Devices (1 Pg)

Exhibit 5-13: Substantially Equivalent Medical Devices (1 Pg)

Operation		Submit 510(k)	Register	List	COMPLY W/GMP
1.	Manufacture and distribute device	YES: 807.81(a)	YES 807.20	YES 807.20(a)	YES
2.	Contract manufacturer who commercially distributes device for specifications developer	NO: 807.81(a)	YES if domestic: 807.20(a)(2), YES if foreign 807.40(a)	YES if domestic 807.20(a)(2), YES if foreign 807.40(a)	YES
3a.	Contract manufacturer who meets the definition of finish-ed device manufacturer per 21 CFR 820.3(l), but does not commercially distribute device for specifications developer	NO	NO: 807.20(c)(1)	NO: 807.20(c)(1)	YES
3b.	Contract manufacturer who does not meet the definition of finished device manufacturer per 21 CFR 820.3(l) (e.g., component manufacturer, subassembler) but does not commercially distribute device for specifications developer	NO	NO: 807.20(c)(1)	NO: 807.20(c)(1)	NO
4	Manufacturer modifies device or new intended use and distribute	NO: preamble no. 17 & 18 FR 8/23/77 YES: 807.81(a)(3) with signif. change in device or use	YES 807.20(a)	YES 807.20(a)	YES
5	Distribute U.S. Made device: no specification initiation (domestic distributor)	NO: 807:85(b)	NO: 510(g)(4) of act, 807.20(c)(3)	NO	NO
6	Specification initiator and distribute only	YES: 807.81(a)	YES: 807.20(a)(1) preamble no. 5, FR 8-23-77	YES: 807.20(a)(1)	YES: 820.181, etc.
7	Specification consultant only; no distribution	NO	NO: preamble no, 5, FR 8-3-77	NO	NO
8	Relabeler or repacker: distribute under own name	NO: 807.85(b): no change to device or existing labeling	YES: 807.20(a)(3)	YES: 807.20(a)(3) preamble no. 7, FR 8-25-78	YES

9	Kit assembler using prelabeled & prepackaged devices only	NO: no change in device or existing labeling other than adding dist. name & address 807.81(a)(3)	YES: 807.20(a)	YES: 807.20(a)	NO
10	Kit assembler changes intended use (801.4) of prepackaged/prelabeled devices	YES: 807.81(a)	YES: 807.20(a)(2)	YES: 807.20(a)(2)	YES: 820.120, 820.130, etc.
11	Kit assembler changes prepackaged/prelabeled devices	NO: if no significant change to labeling or device: otherwise YES: 807.81(a)(3)(i)	YES: 807.20(a)(3)	YES: 807.20(a)(3)	YES
12	Manuf. Accessory, component and package & label for health purpose to end user.	YES: 807.81(a)	YES: 807.20(a)(5) preamble no. 77, FR 8-25-78	YES: 807.20(a)(5)	YES
13	Manuf. Components & dist. Only to finished device mfr.	NO: 807.81(a)	NO: 807.65(a)	NO	Use as guide: 820.1
14	Contract mfr. Of subassembly or component (see no. 12, accessory)	NO	NO	NO	Primary mfr. must see that GMP is met preamble no. 33, FR 7-21-78
15	Contract packager or labeler	NO	NO	NO	Primary mfr. must see that GMP is met preamble no. 33, FR 7-21-79
16	Contract sterilizer who commercially distributes device	NO	YES if domestic 807.20(a)(2), YES if foreign 807.40(a)	YES if domestic 807.20(a)(2), YES if foreign 807.40(a)	YES
17	Contract sterilizer who does not commercially distribute device	NO	NO: 807.20(c)(2)	NO: 807.20(c)(2)	YES
18	Manufacture custom device (domestic or foreign)	NO: 807.85(a)(1)&(2)	YES 807.20(a)(2)	YES 807.20(a)(2)	YES: also see 520(b); 520(f)
19	U.S. Establishment who manufactures for export only	NO	YES 807.20(a)(2)	YES 807.20(a)(2)	YES

Exhibit 5-13: Substantially Equivalent Medical Devices (1 Pg)

20	Foreign manufacturers and all foreign establishments	YES: 807.81 foreign mfr. has primary responsibility, but may delegate to an init. Dist.	YES, 807.40(a)	YES 807.40(a)	YES
21	Initial distributor/importer of device	YES: 807.81(a) or 807.85(b) unless 510(k) has been filed by foreign manufacturer or another init. Dist	YES: 807.40(a)	NO: enforcement discretion used for 807.22(c)	YES: 807.3(d), 820.198, 820.100, 820.200, etc.
22	Installer-mfr.'s agent	NO	NO	NO	YES: 820.170
23	Installer-user	NO	NO	NO	NO: for x-ray see 1020.30(d) report
24	Device being investigated under ide	Exempt: 812.1(a)	NO	NO: 807.40(c)	Exempt per 812.1(a), except for Design Control per 820.30
25	Mfr. Buys manufacturing rights for device (see no. 4)	NO: preamble 18 FR 8-23-77 only if same type of manuf. equip. is used and no signif. change to device	YES: 807.20(a)(2) if not already registered	Send letter to FDA per 807.30(b)(5) & 807.26	YES
26	Reprocessor of single use device	YES	YES: 807.20	YES: 807.20	YES
27	Foreign exporter of device (device manufactured in foreign country)	YES: (original manufacturer's 510(k) maybe used)	YES: 807.40 (a)	YES: 807.40 (a)	

*Includes W, WA, WF, WR, and/or WZ

Exhibit 5-14: Facts - Profile - Comstat (6 Pgs)

This document is available on the FDA website at http://www.fda.gov/ICECI/Inspections/IOM/ucm127372.htm.

Exhibit 5-15: Compliance Achievement Report

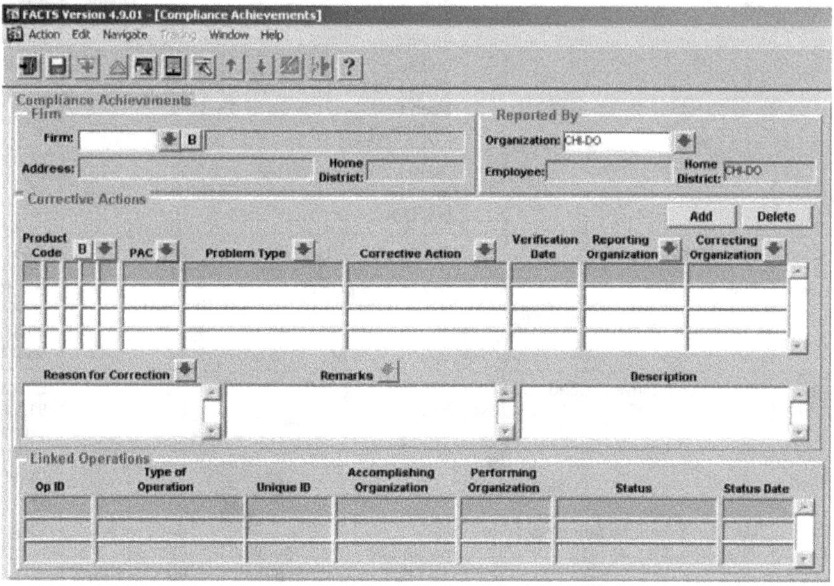

Exhibit 5-16: Facts El Record (3 Pgs)

Exhibit 5-16: Facts El Record (3 Pgs)

FACTS Version 4.9.01 - [Maintain Inspection Results]

Action Edit Options Navigate Tracing Window Help

○ Page 1 ○ Page 2

Inspection Results

FEI: 3000900702 B Name: Standard Seafood Co.
Address: 9056 Telegraph Rd
Milwaukee, Wisconsin, 53204, United States

Accomplishing District: MIN-DO
Start Date: 10/03/2005
Compltn Date: 10/17/2005

Status: In Progress
Inspection Basis: Compliance
Compliance Number:

Inspected Processes & Conclusions

PAC	Establishment Type	Process (Product)	Reschedule Inspection Date	Re-Inspection Priority	Inspection Conclusion	Pro Rqd.	
03803	Manufacturer	16 K F T			Correction Indicated (CI)		Add
							Delete

District Decisions

Final Decision?	Decision Date	Decision Type	Decision Made By	Organization Name	Remarks	
☐						Add
☐						Delete
☐						

Products Covered

Product Code	B	Establishment Type	Description	Additional Product Description	
16 K F T 05		Manufacturer	Shrimp & Prawns, Breaded; Paper; Pa		Add
					Delete

FACTS Version 4.9.01 - [Maintain Inspection Results]

Action Edit Options Navigate Tracing Window Help

○ Page 1 ● Page 2

Inspection Results

☑ 483 Issued?
483 Location: CF Jacket
TRIPS Number:

Endorsement Location: FACTS
EIR Location: CF Jacket
MQSA Status:

Endorsement Text: Previous inspection of 3/04 revealed significant violative conditions with respect to the firm's frozen raw breaded shrimp production and

Inspection Summary: Firm is a manufacturer of frozen raw breaded shrimp and other frozen seafood products, including cooked shrimp and raw clams/

IB Suggested Actions

Action	Remarks	
Citation		Add
		Delete

Referrals

Org Name	Mail Code	Referral Reason	
MIN-CB	HFR-MW340	violative EIR	Add
			Delete

Refusals

Inspection Refusals	
Refusal to permit review of control reco	Add
Refusal to permit review of complaint files	Delete
Refusal to permit photography	

Samples Collected

Sample Number	
2232	Add
	Delete

Recall Numbers

Recall Number	
	Add
	Delete

Related Complaints

Consumer Complaint Number	
	Add
	Delete

Inspection Accomplishment Hours

Operation

Operation Code: 12 - Domestic Inspection   Work Subject / Title: MN Seafood FY 05

Assignment Status: In Progress   Status Date: 07/27/2005   Reimbursable: ☐

Performing Org.: MIN-GRP6

Assignees Accomplishment Hours

Lead Invstgtr	Employee Name	Position Class	Hours Credited To	PAC	Est Type	Process	Hours
☑	Rogers, Sidney H	INV	MIN-DO				
☐							
☐							
☐							
☐							
☐							

Add
Delete
Duplicate

Total Hours :

Inspection Basis

Find   %

Description
Compliance
Consumer Complaint
Surveillance

Find   OK   Cancel

Inspection Refusals

Find   %

Description
Refusal to permit review of underlying data
No refusal
Refusal to permit entry
Refusal to allow inspection except by appointment or other
Refusal to furnish qualitative or quantitative formulae
Refusal to disclose or permit observation of mfr. procedures
Refusal to permit review of control records
Refusal to permit review of complaint files
Refusal to permit review of sales or shipping records
Refusal to permit collection of samples
Refusal to permit photography

Find   OK   Cancel

FDA 483 Responses

483 Responses

Response Type	Response Mode	Response Date	Response Summary
Adequate, Requires Verification	Letter	10/21/2005	

OK
Add
Delete

Exhibit 5-16: Facts EI Record (3 Pgs)

Maintain Products Covered

Products Covered

FEI: 3000900702

Name: Standard Seafood CO.

Address: 9058 Telegraph Rd.

Milwaukee, WI 53204

United States

Discontinued Product ?	Product Code					Establishment Type	Description	Last Covered Date	
☐	16	K	F	T	05	Manufacturer	Shrimp & Prawns, Breaded, Paper, Packaged	10/17/2005	Select
☐	16	J	G	N	05	Manufacturer	Shrimp & Prawns, Plastic, Synth, Heat Treated		Add
☐	16	E	H	T	02	Manufacturer	Clams, Nonflex Plastic, Packaged Food (Not Comme		Delete
☐									
☐									
☐									
☐									
☐									
☐									
☐									
☐									
☐									

Part III

Combined Glossary and Index

Combined Glossary

A, B

Advanced Prepared Food

> Food that was prepared on location at the food service establishment prior to arrival of the Lead Investigator. [Chapter 3]

Bacteriological Samples

> During inspections of firms producing products susceptible to microbial contamination (e.g., frozen precooked; ready to eat seafood, creme filled goods, breaded items, egg rolls, prepared salads, etc.), proof of adulteration, with fecal organisms, or elevated levels of non-pathogenic microorganisms, must be established. Sampling of raw materials, in-line and finished product is warranted. Follow instructions under IOM 4.3.7.7 - Products Susceptible to Contamination with Pathogenic Microorganisms, Sampling During Inspection. [Chapter 4]

Biological Product

> A virus, therapeutic serum, toxin, antitoxin, vaccine, blood, blood component or derivative, allergenic product, or analogous product, or arsphenamine or derivative of arsphenamine (or any other trivalent organic arsenic compound), applicable to the prevention, treatment, or cure of a disease or condition of human beings (Public Health Service Act Sec. 351(i)). Additional interpretation of the statutory language is found in 21 CFR 600.3. Biological products also meet the definition of either a drug or

device under Sections 201(g) and (h) of the Federal Food, Drug, and Cosmetic Act (FD&C Act).

Veterinary biologicals are subject to the animal Virus, Serum, and Toxin Act which is enforced by USDA (21 U.S.C. 151-158). [Chapter 5]

C

Citation (Cite)

The section 305 Notice is a statutory requirement of the FD&C Act. It provides a respondent with an opportunity to show cause why he should not be prosecuted for an alleged violation. Response to the notice may be by letter, personal appearance, or an attorney(s). [Chapter 2]

Civil Number

A docket number used by US district courts to identify civil cases (seizure and injunction). [Chapter 2]

Complaint for Forfeiture

A document furnished to the U.S. attorney for filing with the clerk of the court to initiate a seizure. [Chapter 2]

Complaint Samples, Certain

Injury and illness investigation samples from certain complaints where there is no Federal jurisdiction, or where the alleged violation offers no basis for subsequent regulatory action. Complaint samples from lots for which Federal jurisdiction is clear should be submitted as Official Samples. [Chapter 4]

Comprehensive Inspection

Directs coverage to everything in the firm subject to FDA jurisdiction to determine the firms compliance status. [Chapter 5]

Criminal Number

A docket number used by the US district courts to identify criminal cases (prosecutions). [Chapter 2]

D

Dealer

For sample collection purposes, the dealer is the person, firm (which could include the manufacturer), institution or other party, who has possession of a particular lot of goods. The dealer does not have to be a firm or company, which is in the business of buying or selling goods. The dealer might be a housewife in her home, a physician, or a public agency; these dealers obtain products to use but not to sell. The dealer may be a party who does not own the goods, but has possession of them, such as a public storage warehouse or transportation agency. [Chapter 4]

Directed Inspection

Directs coverage to specific areas to the depth described in the program, assignment, or as instructed by your supervisor. [Chapter 5]

District Contact

The Director, Investigations Branch. [Chapter 3]

Denaturing

Decharacterization of a product, whereby it is made unusable for its originally intended purpose. [Chapter 2]

Destruction

> The procedures involved in rendering a product unsalvageable. Destruction may be accomplished by burning, burial, etc. [Chapter 2]

Device

> Section 201(h) of the FD&C Act [21 U.S.C. 321 (h)] defines a device as follows: "The term "device" *** means an instrument, apparatus, implement, machine, contrivance, implant, in-vitro reagent, or other similar or related article, including any component, part, or accessory, which is: [Chapter 2]

> 1. Recognized in the official National Formulary, or the United States Pharmacopoeia, or any supplement to them,

> 2. Intended for use in the diagnosis of disease or other conditions, or in the cure, mitigation, treatment, or prevention of disease, in man or other animals, or

> 3. Intended to affect the structure or any function of the body of man or other animals, and which does not achieve its primary intended purposes through chemical action within or on the body of man or other animals and which is not dependent upon being metabolized for the achievement of any primary intended purposes."

Document

> Official records which are considered to be U.S. Government property regardless of the media e.g. Regulatory notes (electronic and hardcopy), memoranda, inspection reports, e-mails, and official government forms (e.g. SF-71, FDA-482, FDA-483, etc.) [Chapter 1]

E

Egg and Egg Products (Dual Jurisdiction)

The term "egg" means the shell egg of the domesticated chicken, turkey, duck, goose, or guinea.

The term "egg product" means any dried, frozen, or liquid eggs, with or without added ingredients, excepting products which contain eggs only in relatively small proportion or historically have not been, in the judgment of the Secretary, considered by consumers as products of the egg food industry, and which may be exempted by the Secretary under such conditions as he may prescribe to assure the egg ingredients are not adulterated and such products are not represented as egg products. This would be done on a case by case basis by USDA. [Chapter 2]

Ethylene Oxide (EO)

EO is a colorless gas or volatile liquid with a characteristic ether-like odor above 500 ppm. Unmonitored and inadequate ventilation will allow EO buildup of extremely high concentrations, especially in facilities utilizing malfunctioning or leaking equipment. [Chapter 1]

Exhibits

Filth exhibits and other articles taken for exhibit purposes during inspections to demonstrate manufacturing or storage conditions, employee practices, and the like. Typically filth exhibits submitted as part of an INV sample are not tied to any specific lot of product, but are meant to illustrate the conditions at a firm. An example of an INV filth sample would be rodent excreta pellets, apparent nesting or other rodent gnawed material, and other evidence of rodent activity collected from the perimeter and at multiple locations throughout a manufacturing facility or warehouse in order to document widespread rodent infestation. [Chapter 4]

F

Factory Samples

Raw materials, in-process and finished products to demonstrate manufacturing conditions. Note: Photographs taken in a firm are not samples. They are exhibits except when they are part of a DOC Sample. See IOM 4.5.2.4, 5.3.3, and 5.3.4. [Chapter 4]

FDC and INJ Numbers

The number used by the Chief Counsel's office to identify FDA cases. [Chapter 2]

Food

For the purpose of detention of food under section 304(h) of the FD&C Act, see section 201(f) of the FD&C Act, which defines food as follows: "(1) articles used for food or drink for man or other animals, (2) chewing gum, and (3) articles used for components of any such article."

Examples of food include, but are not limited to, fruits, vegetables, fish, dairy products, eggs, raw agricultural commodities for use as food or components of food, animal feed, including pet food, food and feed ingredients and additives, including substances that migrate into food from food packaging and other articles that contact food, dietary supplements and dietary ingredients, infant formula, beverages, including alcoholic beverages and bottled water, live food animals, bakery goods, snack foods, candy, and canned foods. [Chapter 2]

Food Service Function

A public event where food will be provided to a protectee. [Chapter 3]

G – L

Home District

> The Home District is the district in whose territory the alleged violation of the Act occurs, or in whose territory the firm or individual responsible for the alleged violation is physically located. The original point from which the article was shipped, or offered for shipment, as shown by the interstate records, is usually considered the point where the violation occurred; and the shipper of such article, as shown by such records, may be considered to be the alleged violator.

> Where actions against a firm are based on goods which became violative after interstate shipment was made, or after reaching its destination (such as 301(k) violations), the dealer in whose possession the goods are sampled may be considered the violator, and the location of this dealer determines the "Home District". [Chapter 2]

Lead Advance Agent

> The Secret Service Agent in charge of all security arrangements. This person is responsible for all sites to be visited by the protectee, and is a representative of the Office of Protective Operations (Secret Service Headquarters). [Chapter 3]

Lead Investigator

> The FDA person designated by the FDA district/region to coordinate the investigational activities at the site of a food service function. [Chapter 3]

M

May Proceed

Product may proceed without FDA examination. FDA has made no determination the product complies with all provisions of the Food, Drug, and Cosmetic Act, or other related acts. This message does not preclude action should the products later be found violative." (No compliance decision has been made.) [Chapter 6]

Meat Products and Poultry Products (Dual Jurisdiction)

For FDA purposes, meat products and poultry products are defined as the carcasses of cattle, sheep, swine, goats, horses, mules, other equines, or domesticated birds, parts of such carcasses, and products made wholly or in part from such carcasses, except products exempted by U.S.D.A. because they contain a relatively small amount of meat or poultry products (e.g.; meat flavored sauces, pork and beans, etc.). Examine labels for USDA Shield or coding information to help determine if it is a USDA product. [Chapter 2]

Medical Device Notification

A communication issued by the manufacturer, distributor, or other responsible person in compliance with a Notification Order. It notifies health professionals and other appropriate persons of an unreasonable risk of substantial harm to the public health presented by a device in commercial distribution. [Chapter 7]

Medical Device Notification Order

An order issued by FDA requiring notification under section 518(a) of the FD & C Act [21 U.S.C. 360h (a)]. The directive issues when FDA determines a device in commercial distribution, and intended for human use, presents an unreasonable risk of substantial harm to the public health. The notification is necessary to eliminate the unreasonable risk of such harm, and no more practicable means is available under the provisions of the Act to eliminate such risk. [Chapter 7]

Medical Device Safety Alert

This is a communication voluntarily issued by a manufacturer, distributor, or other responsible person (including FDA). It informs health professionals and other appropriate persons of a situation which may present an unreasonable risk to the public health by a device in commercial distribution. [Chapter 7]

Mold Samples

During inspections of manufacturers such as canneries, bottling plants, milling operations, etc., it may be necessary to collect scrapings or swabs of slime or other material to verify the presence of mold. The sample should represent the conditions observed at the time of collection and consist of sufficient material to confirm and identify mold growth on the equipment. If possible, take photographs and obtain scrapings or bits of suspect material. Describe the area scraped or swabbed, e.g., material was scraped or swabbed from a 2" x 12" area. [Chapter 4]

N, O

Nolle Prosequi (Nol-Pros)

The prosecutor or plaintiff in a legal matter will proceed no further in prosecuting the whole suit or specified counts. [Chapter 2]

Nolo Contendere (Nolo)

A plea by a defendant in a criminal prosecution meaning "I will not contest it". [Chapter 2]

Official Sample

An Official Sample is one taken from a lot for which Federal jurisdiction can be established. If violative, the Official Sample provides a basis for administrative or legal action. Official Samples generally, but not always, consist of a physical portion of the lot sampled. [Chapter 4]

P, Q

Perishable Food

For the purpose of detention of food under section 304(h) of the FD&C Act, the term "perishable food" means food that is not heat-treated; not frozen; and not otherwise preserved in a manner so as to prevent the quality of the food from being adversely affected if held longer than 7 calendar days under normal shipping and storage conditions. See 21 CFR 1.377. [Chapter 2]

Person-in-Charge

The available person in the food service establishment authorized to make necessary changes/decisions such as the general manager, executive chef, banquet manager, caterer's representative or other management person. [Chapter 3]

Pesticide Episode

An "episode" is defined as a violative pesticide (or other chemical contaminant) finding and all samples collected in follow-up to that finding. All samples must be associated with one responsible firm (grower, pesticide applicator, etc.) and one specific time period (e.g. growing season). [Chapter 4]

Pre-prepared Food

Potentially hazardous food that was received at the food service establishment in a prepared form. Examples would include chicken salad, liver pate, gefilte fish, hors d'oeuvres, etc. which were prepared at another location, and then transported to the food service establishment providing food for the event. [Chapter 3]

Protectee

Any person eligible to receive the protection authorized by law. [Chapter 3]

Protective Detail

A team of Secret Service agents responsible for security surrounding public events to be attended by a protectee during a trip. Protective details are assigned and coordinated by Secret Service Headquarters, but may include Secret Service field representatives. [Chapter 3]

R

Recall

A Recall is a firm's removal or correction of a marketed product that FDA considers to be in violation of the laws it administers, and against which the Agency would initiate legal action (e.g., seizure). Recall does not include a market withdrawal or a stock recovery. See the Agency recall policy outlined in 21 CFR 7.1/7.59 - Enforcement Policy - General Provisions, Recalls (Including Product Corrections) - Guidance on Policy, Procedures, and Industry Responsibilities. [Chapter 7]

Recall Audit Check

A personal visit, telephone call, letter, or a combination thereof, to a consignee of a recalling firm, or a user or consumer in the chain of distribution. It is made to verify all consignees at the recall depth specified by the strategy have received notification about the recall and have taken appropriate action. [Chapter 7]

Recall Classification

Means the numerical designation, i.e., I, II, or III, assigned by the FDA to a particular product recall to indicate the relative degree of health hazard presented by the product being recalled. [Chapter 7]

Class I Recall

A situation in which there is a reasonable probability that the use of, or exposure to, a violative product will cause serious adverse health consequences or death.

Class II Recall

A situation in which use of, or exposure to, a violative product may cause temporary or medically reversible adverse health consequences or where the probability of serious adverse health consequences is remote.

Class III Recall

A situation in which use of, or exposure to, a violative product is not likely to cause adverse health consequences.

Recall Completed

For monitoring purposes, the FDA classifies a recall action "Completed" when all outstanding product, which could reasonably be expected is recovered, impounded, or corrected. [Chapter 7]

Recall, Depth of

Depending on the product's degree of hazard and extent of distribution, the recall strategy will specify the level in the distribution chain to which the recall is to extend, i.e., wholesaler, retailer, user/consumer. [Chapter 7]

Recall Number

> Number assigned by a responsible Center for each re-called product they initiate. This number consists first of a letter designating the responsible Center (see letter Codes below), a 3-digit sequential number indicating the number of recalls initiated by that Center during the fiscal year, and a 1-digit number (the Center for Devices and Radiological Health (CDRH) uses 2-digit numbers) indicating the fiscal year the recall was initiated. For example: F-100-2 identifies the 100th recall initiated by the Center for Food Safety and Applied Nutrition (CFSAN) in FY-2002. [Chapter 7]

Recall Strategy

> A planned specific course of action to be taken in conducting a specific recall, which addresses the depth of recall, need for public warnings, and extent of effectiveness checks for the recall. [Chapter 7]

Recall Terminated

> A recall will be terminated when the FDA determines that all reasonable efforts have been made to remove or correct the violative product in accordance with the recall strategy, and when it is reasonable to assume that the product subject to the recall has been removed and proper disposition or correction has been made commensurate with the degree of hazard of the recalled product. Written notification that a recall is terminated will be issued by the appropriate District office to the recalling firm. [Chapter 7]

Recall Type

> A designation based on whether the recall is Voluntary, FDA Requested (at the request of the Commissioner or his designee), or ordered under section 518(e) of the FD & C Act [21 U.S.C 360h (e)]. [Chapter 7]

Reconditioning

> The reworking, relabeling, segregation, or other manipulation which brings a product into compliance

with the law, whether or not for its original intended use. [Chapter 2]

Reconditioning Samples

These are taken from lots reconditioned under a Decree or other agreement to bring the lots into compliance with the law. The sample is taken to determine if reconditioning was satisfactorily performed. These samples should be submitted as Official Samples, rather than INV. [Chapter 4]

Release

The product is released after FDA examination. This message does not constitute assurance the product complies with all provisions of the Food, Drug and Cosmetic Act, or other related Acts, and does not preclude action should the product later be found violative." (A compliance decision has been made.) [Chapter 6]

S

Seizing District

The district where seizure is actually accomplished. The seizing district is not necessarily the collecting district, as in the case of intransit samples. [Chapter 2]

Subpoena Duces Tecum

A writ commanding a person to appear in court bringing with him certain designated documents or things pertinent to the issues of a pending controversy. [Chapter 2]

Supervising District

The district which exercises supervision over reconditioning lots in connection with seizure actions. [Chapter 2]

Seizure

> Seizure is a judicial civil action directed against specific offending goods, in which goods are "arrested." Originally designed to remove violative goods from consumer channels, it was intended primarily as a remedial step; however, the sanction often has a punitive and deterrent effect. [Chapter 2]

Site Advance Agent

> The Secret Service person responsible for security arrangements at a specific site to be visited by the protectee. This person is part of the protective detail headed by the Lead Advance Agent. Note: the term Site Advance Agent will include any agent designated by the Site Advance Agent to be the contact with the FDA Lead Investigator. [Chapter 3]

Support Personnel

> FDA persons deemed necessary by FDA in order to properly inspect a food service function. [Chapter 3]

T – Z

Verified

> To confirm; to establish the truth or accuracy. [Chapter 5]

Voluntary Corrective Action

> The observed voluntary repair, modification, or adjustment of a violative condition, or product. For purposes of this definition, violative means the product or condition does not comply with the Acts or associated regulations enforced by the Agency. [Chapter 2]

Index

703 Record Requests   5.1.1.7.2

A

Access   5.2.9.2.1
Addendum to EIR   5.10.6
Additional Information   5.10.4.3.16
Administrative Data   5.10.4.3.3
Adverse Event Reporting   5.5.7
Animal Grooming Aids   5.9.7
Annotation of the FDA 483   5.2.3.4
Anonymity   5.2.9.1.3
Antibiotic   5.2.3.2
Approved Drugs   5.5.5.5
Attachments   5.10.4.3.20
Authority for Examinations and
    Investigations   5.1.1.12
Authority to Enter and Inspect   5.1.1
Authority to Implement Section 702(E)(5) of
    FD&C Act   5.1.1.13

B

Banned Devices   5.6.8
Basic Premises   5.2.1.1.1
Biological Product   5.7.1
Biologics   5.7
Biologics Inspection   5.7.2
    Approach   5.7.2.5
    Authority   5.7.2.1
Blood and Source Plasma Inspections   5.7.2.1.1
    Brokers   5.7.6
    Core Team   5.7.2

Donor Confidentiality	5.7.2.2
Guidelines	5.7.2.6
Human Tissue Inspections	5.7.2.1.2
Inspectional Objectives	5.7.2.3
Licensing	5.7.3
Listing	5.7.3
Preparation	5.7.2.4
Recommendations	5.7.2.6
Registration	5.7.3
Regulations	5.7.2.6
Responsible Person	5.7.4
Team Biologics	5.7.2
Technical Assistance	5.7.2.7
Testing Laboratories	5.7.5
Bio-research Monitoring – CBER	5.7.2.8
Bio-research Monitoring – CFSAN	5.4.1.3
Bio-research Monitoring – CVM	5.9.7
Bio-research Monitoring – CDER	5.5.1.3
Bio-research Monitoring – CDRH	5.6.1.4
Biosecurity Procedure	
Animal Grower	5.2.10
Animal Husbandry	5.2.10
Animal Producer	5.2.10
Inspection Procedure	5.2.10.2
Pre-inspection Activities	5.2.10.1
Preparation	5.2.10.1
Special Situation	5.2.10.3
Bioterrorism Act	5.4.1.5
Import Requirements under the BT Act	5.1.1.14.1
Records Access	5.4.1.3
Blood and Blood Products Inspection	5.7.2.1.1
Body of Report	5.4.10.2.2
Brokers	5.7.6
BSE Activities	5.9.4
Business Premises	5.1.1.8

C

Candling	5.1.5.1
CBER Bio Research Monitoring	5.7.2.8
CDER Bio Research Monitoring	5.5.1.3

		5.5.6
CDRH Bio Research Monitoring		5.6.1.4
CFSAN Bio Research Monitoring		5.4.1.6
Chlorine Solution Pipes		5.4.5.7
Clinical Investigators and/ or Clinical		
Pharmacologists		5.5.5.7
Clothing		5.1.4.1
Color Additives		5.4.6.4
Color Additives Status List		5.4.6.4
Color Slide Identification		5.3.4.2.2
Complaint Files		5.4.8.3
Complaint Files		5.6.2.4
Complaints		5.10.4.3.11
Compliance Achievement Reporting System		
(CARS)		5.10.2.1
Computerized Complaint and Failure Data		5.3.8.4.1
COMSTAT		5.10.2
		Exhibit 5-14
Concurrent Administrative, Civil, and Criminal		
Actions		5.2.2.8
Condition		5.4.4.2
Conducting Regulatory Inspections when the		
Agency is Contemplating/Taking,		
Criminal Action		5.2.2.4
Confidentiality		5.1.4.3
Confiscation		5.1.1.13
Consumer Complaints		5.2.8
Contract Facilities		5.2.3.2
		5.6.6
Conveyor Belt Condtions		5.4.5.3
Copies		5.2.3.6.2
Correction of FDA 483 Errors		5.2.3.1.6
Counterfeit Drug		5.1.1.13
Credentials		5.1.1.2
		5.2.2
Criminal Action		5.2.2.4
Criteria for Consideration		5.2.1.1.2
Eligibility Criteria		5.2.1.1.2.2
Type of Inspection		5.2.1.1.2.1
Current Practices		5.8.2
CVM Bio Research Monitoring		5.9.8

CVM Website   5.9.1

D

Data Elements   5.10.2.1.2
Data Integrity of Records Provided By Firm 5.3.8.4.4
Date Issued   5.2.3.1.3
Depth of Inspection   5.1.2.1
Destruction   5.2.9.2.4
Device Inspection
    Authority   5.6.1
    Banned Device   5.6.8
    Complaint Files   5.6.2.4
    Contract Facilities   5.6.6
    Device Inspection Report   5.6.9
    GWQAP   5.6.5
    In-Vitro Diagnostics   5.6.4
                                                        5.7.2
    Labeling   5.6.4
    Medical Device Quality System/GMPs   5.6.2
    Preparation   5.6.2.1
    Quality Audit   5.6.2.2
    Records   5.6.2.3
    Sampling   5.6.1.2
    Small Manufacturer   5.6.7
    Sterile Devices   5.6.3
    Technical Assistance   5.6.1.1
    Types   5.6.1.3
Devices   5.6
Video Recordings   5.3.4.2.4
Digital Photos - Turbo EIR   5.3.4.4
    Resize Photo   Exhibit 5-6
    Insert Photo   Exhibit 5-7
    Resize using MS Office Picture Manager Exhibit 5-8
Directed Inspection   5.1.2
Disclosure   5.2.9.2.3
Discovery of Criminal Violation   5.2.2.5
Discussion on Duty, Power, Responsibility   5.3.6.1
Discussion with Management   5.2.7
                                                        5.10.4.3.15
Distribution   5.4.8

Distribution of the FDA 483	5.2.3.6
Drug Approval Status	5.5.5.2
Drug Inspection	
Adverse Event Reporting	5.5.7
Advertising	5.5.3
Approach	5.5.1.2
Authority	5.5.1
Bioresearch Monitoring	5.5.6
Dietary Supplement Status	5.5.5.4
Drug Inspection Report	5.5.8
General Elements	5.5.8
Guarantee	5.5.4
Inspection References	5.5.1.1
Labeling Agreement	5.5.4
Listing	5.5.2
Preparation	5.5.1.1
Promotion	5.5.3
Registration	5.5.2
Drug Registration & Listing	5.5.2
Drug Status Questions	5.5.5.3
Drug/Dietary Supplement Status	5.5.5.4
Drugs	5.5

E

EIR	5.10.1
Team inspections	5.1.2.5
Electronic Information	5.10.5.1
Electronic Information for Official	
Documentation	5.3.8.4.5
Electronic Products	5.1.1.10
Employee Practices	5.4.7.2.2
Endorsement	5.10.2
Equipment and Utensils	5.4.5
Chlorine Solution Pipes	5.4.5.7
Conveyor Belt Condtions	5.4.5.3
Filtering Systems	5.4.5.1
Mercury and Glass Contamination	5.4.5.5
Sanitation of Machinery	5.4.5.2
Sanitation Practices	5.4.5.8
Utinsels	5.4.5.4

UV Lamps	5.4.5.6
Errors Discovered after Leaving	
Establishment	5.2.3.1.6.2
Errors Discovered Prior to Leaving the	
Establishment	5.2.3.1.6.1
Establish Motivation	5.2.9.1.2
Establishment Inspection Report (EIR) see	
Inspection Report	
Evidence	
Digital Photographs as Regulatory	
Evidence	5.3.4.3
Digital Photographs or Video Recordings	5.3.4.2.5
Digital Photos - Turbo EIR	5.3.4.4
Documentation of Responsibility	5.3.6.2
Duty	5.3.6.1
Evidence Development	5.3
Exhibits	5.3.3
Factory Sample	5.3.2
Guarantee	5.3.7
In-plant Photograph	5.3.4.1
Labeling Agreement	5.3.7.2
Photocopy	5.3.4
Photograph	5.3.4
Power	5.3.6.1
Preparation of Photographs	5.3.4.2
Providing Copies of Photographs	5.3.4.5
Recording	5.3.5
Responsibility	5.3.6.1
Responsible Person	5.3.6
Taping	5.3.5
Evidence Development	5.3
Complaint	5.3.8.4.1
Complaints	5.3.8.4
Computerized Data	5.3.8.4
	5.3.8.4.1
	5.3.8.4.2
Documentary sample	5.3.8.2
Documents	5.3.8
Electronic Data as Official Documentation	5.3.8.4.5
Electronic Records	5.3.8.3
Failure	5.3.8.4

	5.3.8.4.1
Film	5.3.8.3
Identification and Security of Electronic	
Data	5.3.8.4.3
Identification of Records	5.3.8.1
Integrity of Data	5.3.8.4.4
Managing Records Collected	5.3.8.5
Microfiche	5.3.8.3
Microfilm	5.3.8.3
Official Sample	5.3.8.2
Original Records	5.3.8.2
Patient Information	5.3.8.6
Private Information	5.3.8.6
Records	5.3.8
Sampling Request	5.3.9
Evidence Gathered in a Criminal	
Investigation	5.2.2.6
Evidence Voluntarily Provided to the Agency	5.2.2.7
Examination	5.1.1.12
Exemption	5.4.1.5.1
Exemption Requirements	5.3.7.3
Exhibits	5.3.3
	5.10.5
Exhibits Collected	5.10.4.3.19
External Observers	5.1.4.3

F

Facilities Exempted From Registration	5.4.1.5.1
Facilities Where Electronic Products Are Used	
or Held	5.1.1.10
Factory Samples	5.3.2
FACTS Assignment Section	5.3.9.1
FACTS Establishment Inspection Record	
(EI Record	5.10.3
FACTS Operations Section	5.3.9.2
FACTS Organizations Section	5.3.9.3
FACTS Reporting	
Profile COMSTAT	Exhibit 5-13
FDA Investigator's Responsibility	5.1.1.1
Field Exams	5.1.5.3

Filmed or Electronic Records	5.3.8.3
Filtering Systems	5.4.5.1
Firm's Training Program	5.10.4.3.8
Follow Up Guidance	5.1.1.13.3
Follow Up Inspections by Court Order	5.2.2.3
Food	5.4
Food Additives	5.4.6.3
Food Additive Nomographs	Exhibit 5-11
Food and Cosmetic Defense Inspectional	
Activities	5.4.1.4
Food and Cosmetic Security	5.4.1.4.1
Food Chemicals Codex	5.4.4.3
Food Establishment Inspection	5.4.10.1
Food Inspection	
Authority	5.4.1.2
Chemical Codex	5.4.4.3
Color Additives	5.4.6.4
Complaint File	5.4.8.3
Design	5.4.3.1
Distribution	5.4.8
Employees	5.4.2
Entry Review	5.4.1.4
Environment	5.4.3
Equipment	5.4.5
Facilities	5.4.3
Field Examination	5.4.1.4
Food Additives	5.4.6.3
Food Inspection Report	5.4.10.2
Food Standard Inspection	5.4.10.1
Food Standards	5.4.10
Food Transport Vehicle	5.4.7.3.1
Formulas	5.4.6.2
Grade A Dairy Plant Inspection	5.4.9.3
Ingredient Handling	5.4.6.1
Interstate Shipping	5.4.8
Management	5.4.2
Manufacturing Code	5.4.6.5.3
Manufacturing Process	5.4.6
Microbiological Concerns	5.4.7.2
Other Government Inspection	5.4.9
Packaging and Labeling	5.4.6.6

Personnel	5.4.2
Plant Construction	5.4.3.1
Plant Services	5.4.3.3
Preparation	5.4.1.1
Quality Control	5.4.6.5
Raw Material Handling	5.4.4.1
Raw Material Source	5.4.4
Recall Procedure	5.4.8.2
Receiver Vehicle	5.4.7.3.2
Reconciliation Examination	5.4.1.4.2
Routes of Contamination	5.4.7.1
Safety Precautions	5.4.1.4.5
Sanitation	5.4.7
Security Inspection Activities	5.4.1.4.1
Shipper Vehicle	5.4.7.3.3
Storage	5.4.7.3
Violative Inspections	5.4.10.3
Waste Disposal	5.4.3.2
Written Demand for Records	5.4.1.2.1
Written Request for Information	5.4.1.2.2
Food Inspection Report	5.4.10.2
Food Inspections	5.4.1
	2.2.1.2
Food Products	3.4.2.3
Food Recalls	7.2.2
Food Registration	5.4.1.5
Food Safety and Inspection Service	3.2.1.8
Food Sanitation	6.4.3.1
Food Standards	5.4.10
Food Standards Sample	4.1.5
Food Standards, (FS)	4.4.10.2.5
Food Transport Vehicles	5.4.7.3.1
Foodborne Disease	4.3.5.1
Foodborne Outbreaks	8.3
Additional Case History Interviews	8.3.4.3
Analysis of Data	8.3.5
Assistance	8.3.4.2
Attack Rate Table	8.3.5.4
Exhibit 8-8	
Classification of Illness	Exhibit 8-6
Cooperation with Other Agencies	8.3.1.3

Determination	8.3.4.1
Epidemic Curve	8.3.5.1
Epidemiological Investigative Technique	8.3.1
Establishment Investigation	8.3.4.4
Evaluating Epidemiological Data	8.3.4
Follow-Up Guidance	8.3.2
Contacting the Complainant	8.3.2.2.1
Information to Gather	8.3.2.2.3
Interviews	8.3.2.2
Medical Records	8.3.2.3
Preparation	8.3.2.1
Setting Communication Level	8.3.2.2.2
Foreign Flag Vessels	8.3.1.1
Incubation Periods	8.3.5.3
Interstate Conveyances	8.3.1.2
Interviews	8.3.4.5
Outbreaks	8.3.1
Pathogen Growth Factors	8.3.4.7
Possible Contamination Source	8.3.4.6
Pests	8.3.4.6.1
Poor Sanitation	8.3.4.6.3
Raw Meat	8.3.4.6.2
Workers	8.3.4.6.4
References	8.3.7
Salmonella Enteritidis (SE) in Eggs	8.3.1.4
Sample Handling	8.3.3.3
Sample Size	8.3.3.2
Sampling	8.3.3.1
Symptoms Determination	8.3.5.2
Tracebacks of Foods	8.3.5.5
Foods Rejected by USDA	3.2.1.1
Foods, Dietary Supplements & Cosmetics Injury or Reaction	8.4.5
Foreign Firms	5.1.3
Foreign Trade Zones	6.7.18
Formal Entries	6.2.3.1
Formal Entry	6.7.17
Format for Regulatory Notes	2.1.4
Forms and Other Publications	1.10.2.7
Formula/Label Correction	2.6.4.2.6
Formulas	5.4.6.2

Fourth amendment   5.2.2.4
Free Flowing Liquids   4.3.8.2.1
Freedom of Information Act   1.4.4
    Complainant Access to Report/ Results   8.2.5.4
    Procedures   1.4.4.1
    Request for Documents   1.4.4.2
Freight Bill   4.4.7.2.4
Frequent Flyer Miles   1.2.1.1
Frozen Samples   4.5.3.5
Fumigated   4.4.10.1.6
Fumigation   4.5.3.1
Fumigation Safety Precautions   4.5.3.1.1

G

Gainsharing   1.2.1.5
General Considerations for All Affidavits   4.4.8.1
General Discussion with Management   5.10.4.3.15
General Inspection Procedures   5.2.10.2
General Investigation Reporting   8.10
General Procedures (aseptic sampling)   4.3.6.1
General Procedures (investigations)   8.8.5.1
General Procedures & Techniques   5.1.5
Glossary of Digital Terminology   5.3.4.2.6
    Digital Data   5.3.4.2.6.1
    Analog Data   5.3.4.2.6.2
    Memory Card   5.3.4.2.6.3
    Original   5.3.4.2.6.4
    Original Copy   5.3.4.2.6.5
    Permanent Storage Media   5.3.4.2.6.6
    Time/Date Stamp   5.3.4.2.6.7
    Working Copy   5.3.4.2.6.8
Glossary of Import Terms   6.7
Government Agency   4.2.7
Government Bill of Lading   4.5.5.8.3
Government Furnished Vehicles   1.2.2
Government Wide Quality Assurance Program   5.2.3.5
                                                5.6.5
GovTrip   1.2
Grade A Dairy Plant Inspections   5.4.9.3
Grain Elevators   1.5.3.3.2

Grand Jury	5.2.2.9
Grand Jury Proceedings	2.2.7.3
Grower	4.4.10.3.26.2
Growers	5.8.3
Guarantees and Labeling Agreements	5.3.7
Guarantees and Labeling Agreements	5.5.4
Guaranty	5.3.7.1
GWQAP Samples	4.1.6

H

Hand Ship	6.5.5.7
Handling Procedure	5.4.4.1
Hantavirus Associated Diseases	1.5.5.4
Harvester	4.4.10.3.26.3
Hazard Analysis Critical Control Points (HACCP)	4.3.7.7
Hazardous Waste Sites	8.5.5.6
Headquarters	2.2.6.2
Health Fraud	8.6.1
Health and Hygiene	1.5.1.5
Health Care Financing Administration (HCFA)	3.2.4.4
Health Services Administration (HSA)	3.2.4.5
Hearing for Injunction	2.2.8.2
Hearing Protection	1.5.1.2
HHS MOU's	3.2.4.1
History	5.10.4.3.4
History of Menu Items	Exhibit 3-2
Home District	2.2.5.6
Hospitalized In-Travel Status	
Per Diem Coverage	1.2.4.2
Hostile and Uncooperative Interviewees	5.2.5.4
Hours	4.4.10.3.29
How Prepared	4.4.10.3.30
How to Handle the First Contact	5.2.9.1
Human Blood & Blood Products	2.9.3.1.1
Human Cells, Tissues, and Cellular and Tissue	
Based Products (HCT/Ps)	2.9.3.1.2
Donor Confidentiality	5.7.2.2
For Transplantation, Infusion, or Transfer	7.2.5
Inspections	5.7.2.1.2
Registration and Listing	5.7.3.1.1

Hurricanes	8.5.5.4

I

Identification	4.5.2
	4.5.2.3
Identification and Security of Electronic Storage	
Media	5.3.8.4.3
Identification of Documentation	4.4.5
Identification of lots and records	4.2.6
Identification of Records	5.3.8.1
Identification Techniques	4.5.2.3
Identifying Lot(s) Sampled	4.3.2.2
Identifying Marks	4.5.2
Identifying Original Paper Records	5.3.8.2
IFE Entry Review	6.2.3.4.1
Immediate Delivery (ID) /	
Conditional Release	6.7.19
Immediate Transportation (IT	6.7.26
Import for Export	5.1.1.14
Import for Export (IFE) Entries	6.2.3.4
Export Reform and Enhancement Act	6.2.3.4
Record-keeping requirement	6.2.3.4
IFE Entry Review	6.2.3.4.1
Affirmation of Compliance	6.2.3.4.1
Domestic Follow-up	6.2.3.4.2
Inspection Guidance	6.2.3.4.3
Import Glossary of terms	
American Goods Returned	6.7.1
Bonded Warehouse	6.7.2
Break-bulk Cargo	6.7.3
Consumption Entry (CE)	6.7.4
Container Freight Station (CFS	6.7.5
Date Collected	6.7.6
Date of Arrival	6.7.7
Date of Availability	6.7.8
Detention	6.7.9
Detention without Physical	
Examination (DWPE)	6.7.10
Domestic Import (DI) Sample	6.7.11
Entry	6.7.12

Entry Admissibility File	6.7.13
Entry Documents	
(Entry Package)	6.7.14
Failure To Hold	6.7.15
Filer	6.7.16
Foreign Trade Zones	6.7.17
Formal Entry	6.7.18
Immediate Delivery (ID) /	
Conditional Release	6.7.19
Import Alerts	6.7.20
Importer of Record	6.7.21
Import Sections	6.7.22
Import Status	6.7.23
Importer Misdeclaration	6.7.24
Informal Entry	6.7.25
Immediate Transportion (IT)	6.7.26
Line (Line Item)	6.7.27
Lot	6.7.28
Marks	6.7.29
Port (Point) of Entry	6.7.30
Redelivery Bond	
(AKA Entry Bond)	6.7.31
Stripping (of Containers)	6.7.32
Substitution	6.7.33
Supervisory Charges	6.7.34
Warehouse entry (WE)	6.7.35
Import Investigations	6.1.2
Import Investigation Affidavit	Exhibit 6-5
Import Procedures	
Import Forms Sampling	6.5.4
No Sample/No Examination	6.2.5
No Violation	6.2.6
Post-hearing	
Cost of Supervision	6.2.7.9
	Exhibit 6-3
Destruction	6.2.7.8
Exportation	6.2.7.8
Exportation	6.2.7.10
Notice of FDA Action	6.2.3.6.2
	Exhibit 6-1
Notice of Refusal	

of Admission	6.2.7.8
Notice of Release	6.2.7.7
Violation	
Application for Authorization	
to Relabel or Perform	
Other Acts	6.2.7.3
	Exhibit 6-2
Bond	6.2.7.3
Follow-Up Inspection	6.2.7.4
Importer's Certificate	6.2.7.4
Notice of Detention and Hearing	6.2.7.1
Notice of Refusal of Admission	6.2.7.5
Notice of Refusal of Admission	6.2.7.6
Notice of Release	6.2.7.5
Reconditioning	6.2.7.3
Response	6.2.7.2
Import Sample	4.1.4.9
Import Sample Charts	
Coffee & Dates	Smpl Schdl 8
Whitefish	Smpl Schdl 5
Salmonella	Smpl Schdl 1
In-Line Samples	4.3.7.6
In-Line Sampling	4.3.7.7.1
In-Plant Photographs	5.3.4.1
In-Transit Lots	4.2.5.1
In-Transit Sample	4.1.4.3
	4.3 .4
Examination without a Warrant	4.3.4.1
Examination with a Warrant	4.3.4.2
Resealing Conveyances	4.3.4.3
In-Transit Sampling	4.2.4.3
In-Transit Sampling Affidavit	4.4.7.5
In-Vitro Diagnostic Devices	8.4.3.3
In-Vitro Diagnostics	8.4.3.4.2
Inadequate Prior Notice Submission	6.2.3.5.5
Incubation Periods	8.3.5.3
Indicators	5.2.5.4.1
Individual Headings	5.2.3.1.1
Individual Narrative Headings	5.10.4.3
Individual Responsibility and Persons Interviewed	5.10.4.3.7
Induced Samples	4.1.4.5

	4.3.5.4
Ineffective Recalls	7.3.2.6
Infant Formula	2.9.5.5
Infant Formula and Baby Food	8.2.2
Informal Entries	6.2.3.3
Information Disclosure	
Requests by the public, including trade	1.4.2
Disclosure of official information Privacy Act	1.4
FOIA	1.4
Freedom of Information Act: disclosure	1.4.4
Internal FDA Documents: disclosure	1.4.5
Sharing non-public information with other	
Government officials	1.4.3
Subpoena	1.4.1
Information Exchange and Coordination	3.2.4.3.2
Ingredient Handling	5.4.6.1
Ingredient Supplier	4.4.10.3.26.4
Injunction	2.2.8
Consent Decree	2.2.8.3
District Follow-up	2.2.8.6
Hearing for Injunction	2.2.8.2
Permanent Injunction	2.2.8.5
Preliminary Injunction	2.2.8.5
Temporary Restraining Order	2.2.8.1
Trial	2.2.8.4
Injunction or Criminal Prosecution	4.4.6.2
Injury and Adverse Reaction	
Biologics Injury or Illness	8.4.4.2
CFSAN Regulated Products	8.4.5.3
Cosmetics Injury or Reaction	8.4.5.1
Devices for Implant	8.4.3.2
Drug Injury or Illness	8.4.2
In-Vitro Diagnostic Devices	8.4.3.3
Investigation Procedure	8.4.1
Mechanical or Electromechanical Devices	8.4.3.1
Medical Device Injury or illness	8.4.3.4
Reporting	8.4.8
Biologics Injury/Adverse Reaction Reports	8.4.8.2.7
Drugs	8.4.8.2.1
Foods and Cosmetics	8.4.8.2.3

Licensed Biological Products	8.4.8.2.5
Medical Device and Radiological Products	8.4.8.2.2
Reporting Forms	8.4.8.1
Routing Reports	8.4.8.2
Unlicensed Biological Products	8.4.8.2.6
Veterinary Products	8.4.8.2.4
Sampling	8.4.7
Biological Samples	8.4.7.2
Cosmetic Samples	8.4.7.3
Consumer Complaints	8.4.7.4
Device Samples	8.4.7.1
Vaccine Adverse Event Reporting system	8.4.4.1
Vaccine Adverse Reactions	8.4.4.1
Veterinary Products	8.4.6
Injury Illness	4.3.5.1
Injury/Illness Complaints	8.2.5.2
	8.2.1.1
Injury Samples	4.4.6.3
Insect Contamination	4.3.7.4.3
Collecting Exhibits and/or Subsamples	4.3.7.4.3.2
Examination/Documentation of Contamination	4.3.7.4.3.1
Summary of Sample for Evidence	4.3.7.4.3.3
Insects	5.4.7.1.1
	Appendix A
Inspection after Completion of Authorization to	
Bring Article into Compliance	6.2.7.4
Inspection Guidance	6.2.3.4.3
Inspection Information	5.1
Inspection of Foreign Firms	5.1.3
Inspection of Vehicles	5.2.2.2
Inspection Procedures	5.2
Inspection Procedures	7.2.1
Inspection Refusal	5.2.5
Inspection of Vehicles	5.2.2.2
Inspection Report	
Abbreviated Inspection Report	5.10.4.3
Addendum	5.10.6
Additional Information	5.10.4.3.16
Administrative Data	5.10.4.3.3
Attachment	5.10.4.3.20
Complaint	5.10.4.3.11

Compliance Achievement Reporting System (CARS)	5.10.2.1
Discussion with Management	5.10.4.3.13.2
	5.10.4.3.15
EI Record	5.10.3
EIR	5.10.1
EIR Timeframes	5.10.4.2
Electronic information	5.10.5.1
Endorsement	5.10.2
Establishment Inspection Report	5.10.1
Exhibit	5.10.4.3.19
Exhibits	5.10.5
FACTS Establishment Inspection Record	5.10.3
Comstat Screen	Exhibit 5-15
Maintain Inspection Results Screens	Exhibit 5-16
History	5.10.4.3.4
Interstate Commerce	5.10.4.3.5
Jurisdiction	5.10.4.3.6
Manufacturing Code	5.10.4.3.10
Manufacturing Operation	5.10.4.3.9
Narrative Headings	5.10.4.3
Narrative Report	5.10.4
Non-Violative Establishments	5.10.4.1
Objectionable Conditions	5.10.4.3.13
Recall Procedures	5.10.4.3.12
Refusal	5.10.4.3.14
Responsibility	5.10.4.3.7
Samples Collected	5.10.4.3.17
Signature	5.10.4.3.21
Standard Narrative Report	5.10.4.3.1
Summary	5.10.4.3.2
Summary of Findings report	5.10.4.1
Supporting Evidence	5.10.4.3.13.1
Training Program	5.10.4.3.8
Turbo EIR usage	5.10.4
Voluntary Correction	5.10.4.3.18
Violative Establishments	5.10.4.2
Inspection System	5.4.6.5.1
Inspection Techniques How to Document Responsibility	5.3.6.2
Inspection walk through	5.1.2.2

Inspection Warrant	5.2.6
Inspectional Approach	5.1.2
	5.5.1.2
	5.7.2.5
Inspectional Authority	5.4.1.2
Inspectional Guidance	5.1.1.13.2
	5.4.1.5.3
Inspectional Observations	
Adulteration Observations	5.2.3.2.1
Annotation	5.2.3
Annotation	5.2.3.4
Correction FDA 483 Errors	5.2.3.1.6
Discussion	5.2.3
Distribution	5.2.3.6
FDA 483	5.2.3
	Exhibit 5-5
FDA 483 Statements	5.2.3.1.4
GWQAP	5.2.3.5
Other Observations	5.2.3.2.2
Non-Reportable Observations	5.2.3.3
Preparation FDA 483	5.2.3.1
Reportable Observations	5.2.3.2
Signature	5.2.3
Turbo EIR	5.2.3
Inspectional Precautions	5.1.4
Inspectional Procedure	2.7.2
Inspections	1.5.4
Inspections to Monitor Recall Progress	7.3.1
Inspector/Investigator Role	6.1.3
Intended Use	5.5.5.1
Interaction with Federal Agencies	3.2
Interagency Cooperation	3.2.5.2.6
Interagency Motor Pool	1.2.2.1
Interdistrict Assignments	1.7
Internal FDA Documents	1.4.5
Internal Revenue Service (IRS)	3.2.8.2
International Agreements	3.4
International Inspection	3.1.1
International	
Food Products	3.4.2.3
Memorandum of Understanding	3.4.1

MRA	3.4.2
Mutual Recognition agreement	3.4.2
Pharmaceuticals and Medical Devices	3.4.2.2
Internet	1.10.2.3
Interrogation: Advice of Rights	Exhibit 2-1
Interviewing Informant	5.2.9
Interviewing Persons under Arrest	2.2.11.2
Interviews	8.2.5
Additional Information to Obtain	8.2.5.3
Basic Information to Obtain	8.2.5.1
Complainant Access to Report/Results	8.2.5.4
Injury/Illness Complaints	8.2.5.2
Internet and Intranet	1.10.2.3
Interstate Commerce	5.10.4.3.5
Interstate Commerce	4.4.6.2.1
Interstate Certified Shellfish Shippers	2.9.5.6
Interstate Milk Shippers	7.2.2.1
Interstate Milk Shippers	2.9.5.7
Interstate Shellfish Sanitation Conference	3.5.3
Interviewing Confidential Informants	5.2.9
Interviewing Methods/Techniques	5.2.9.1.1
Interviewing Persons Under Arrest	2.2.11.2
Interviews	3.2.4.3.1
Interviews	8.2.5
Interviews	8.8.5.2
Introduction into I.S.	4.4.6.2.1
Inv. Samples of Filth Exhibits	4.4.10.1.7
Investigation	5.1.1.12
	8.8.5
Investigation, Definition	8.1
Investigation Injury & Adverse Reaction	8.4
Investigation of Foodborne Outbreaks	8.3
Investigation Requirements for Serious Adverse Events of CFSAN Regulated Products	8.4.5.3
Investigation/Reporting	8.4.4.2
Investigational (Inv	4.4.10.2.6
Investigational Device Exemption (IDE) Regulation	2.9.2.2
Investigational Drugs	5.5.5.6
Investigational New Drug Application (IND)	2.9.1.2
Investigational Research	
Data Reporting	8.7.7

Joint Research Project	8.7.2
Priority	8.7.6
Research Assignment	8.7.1
Research Project Identification Code	8.7.3
Research Project Progressive Report	8.7.4
Termination of Research Project	8.7.5
Investigational Samples	4.1.7
Investigations	8.1
Investigations	8.4.1
Investigative Procedures	8.4.2.1
Investigative Procedures	8.4.3.4
Investigator Training and Certification	1.10.2.1
Invoice/Shipping Record FDA 1662	4.4.7.1
	Exhibit 4-8
Ionizing Radiation	1.5.4.2.3
Issuance Authority	1.7.1
Issuance of Detention Termination Notice	
FDA 2291	2.7.2.5.2
Items Not Reported In FACTS	2.6.4.2.9
Items Not Requiring Receipt	5.2.4.2
Items Requiring Receipt	5.2.4.1
Itineraries	1.2.9

J, K, L

Joint Inspections	3.3.1.2
Joint Research Projects	8.7.2
Jurisdiction	
Products Manufactured and/or Distributed	5.10.4.3.6
USDA-FDA Jurisdiction Chart	Exhibit 3-1
LACF / AF Inspections	5.1.1.7.1
Label	4.4.9
Label Review	4.5.3.2
	5.1.5.2
Labeling	6.4.4.1
	5.4.6.6.2
	5.6.4
	4.3.8.3
	4.5.3.2
Labeling Agreement	5.3.7.2
Labels & Accompanying Labeling	4.4.9.1

Labels and Labeling   4.4.9
Laboratory Registration   5.7.3.1.2
Laboratory Tests   5.4.6.5.2
Language Requirements for FDA Documents   1.1
Law, Regulation and Guidance   1.10.1
Laws, Codes, Agencies   3.1.2
Lead Investigator Qualifications   3.2.5.2.8
Leave   1.3
Level of Audit Checks   7.3.2.2
Liability   1.2.2.3
Liaison   3.2.5.2.1
Liaison Officers   3.2.4.3.5
Liaison with Law Enforcement / Intelligence
    Community   8.9.1.2
Limitations   2.2.1.4
Line (Line Item   6.7.27
Listing of Records   5.3.8.5
Living Quarters   5.1.1.9
Locating firms   3.2.16
Lost or Stolen Credentials, Badge   1.6.3.3
Lost or Stolen Equipment   1.6.2.2
Lot   6.7.28
Lot Restoration & Identification   4.3.2
Lot Size   4.4.10.3.31
LACF/AF Food Canning Establishment
    Registration   2.9.5.1

M

Mail Entry   4.4.10.2.7
Mail Entry Sample   4.1.7.2
Mail or Parcel Service Shipments   4.4.7.3
Mail/Personal Baggage   6.2.3.3
Maintenance of Equipment   1.6.2.1
Mammography Quality Standards Act of 1992   2.2.3.8
Man Lifts and Ladders   1.5.4.1
Manufacture within A Territory   4.4.6.2.4
Manufacturer   4.4.10.3.26.5
Manufacturer and Distribution System Follow Up 8.8.5.6
Manufacturer's Raw Materials   2.6.4.2.3
Manufacturing Code System   5.4.6.5.3

Manufacturing Codes	5.10.4.3.10
	4.4.10.3.32
Manufacturing Process	5.4.6
Manufacturing Sites	8.8.5.6.1
Manufacturing/Design Operations	5.10.4.3.9
Map (ORA)	Appendix E
Marks (Imports	6.7.29
Mass media (Press, Radio, and TV)	1.6.1
Meat and Poultry Products	2.7.2.1.1
Meat Products and Poultry Products	2.7.1.3.4
Mechanical, Electrical or Electromechanical	
Devices	8.4.3.1
Medical Device and Radiological Products	8.4.8.2.2
Medical Device Inspections	5.2.3.1.5
Medical Device Notification	7.1.1.8
Medical Device Notification Order	7.1.1.7
Medical Device Quality System/Good	
Manufacturing Practices	5.6.2
Medical Device Recalls	7.2.3
Medical Device Reporting	2.9.2.7
Medical Device Safety Alert	7.1.1.9
Medical Device Samples	4.3.3.1
	8.4.7.1
Medical Records	8.2.6
	8.3.2.3
Medical Record Disclosure FDA 461	Exhibit 8-5
Medicated Feed Mill License (FML)	2.9.4.2
Medicated Feeds and Type A Articles	5.9.3
MedWatch Form	Exhibit 8-10
Memorandum of Understanding	3.1.2.1
Mercury and Glass Contamination	5.4.5.5
Metal Seals	4.5.4.6
Method of Collection	4.4.10.3.33
Method of Payment	4.2.8.3
Method of Shipment	4.5.5.6
Microbiological Concerns	5.4.7.2
Employee Practices	5.4.7.2.2
Processing Equipment	5.4.7.2.1
Microbiological Hazards	1.5.5
Military Blood Banks	5.7.3.1.3
Military Personnel & Civilian Employees' Claims	1.2.2.3.1

Misbranding	4.3.7
	4.4.6.2.2
Mold Contamination	4.3.7.4.6
Moldy Food	2.8.2.2
Monitoring Recalls	7.3
MOU	3.1.2.1
Multiple Date Inspections	5.2.2.1
Multiple FDA 482	5.1.1.11
Multiple Occupancy Inspections	5.1.1.11
Mutual Recognition Agreements	3.4.2
Mycotoxin Sample Chart	Smpl Schdl 6

N

Narcotic and Controlled Rx Drugs	4.2.5.3
Narrative Report	5.10.4
National Center for Drug Analysis	4.5.5.3.1
National Center for Health Statistics	3.2.4.6
National Conference on Interstate Milk Shipments	3.5.2
National Drug Code	4.4.10.3.34
National Institute of Drug Abuse	3.2.4.7
National Institutes of Health (NIH)	3.2.4.8
National Oceanic and Atmospheric Administration & National Marine Fisheries Service	3.2.2.2
National Sample Distributor (NSD)	4.4.10.4
NSD and Assignments	4.4.10.4.1
Overriding NSD	4.4.10.4.2
Other Information	4.4.10.4.3
Natural Disasters	4.3.5.3
Negative Identification	5.3.4.2.3
Net Weight	4.3.8.1
New Animal Drug Application (NADA)	2.9.4.4
New Drug Application (NDA)	2.9.1.3
Nolle Prosequi (Nol Pros)	2.2.5.7
Nolo Contendere (Nolo)	2.2.5.8
Non Government Agreements	3.5
Non Government Meetings	1.6.1.1
Non Injury/Illness Complaints	8.2.1.2
Non Regulatory	4.4.10.2.8
Non Regulatory Sample	4.1.7.3
Non Reportable Observations	5.2.3.3

Non Violative Establishments		5.10.4.1
Notice of Detention & Hearing		6.2.7.1
Notice of Inspection		5.1.1.3
		5.1.2.5
		5.2.2
		Exhibit 5-1
	Carrier	4.1.1.2
	Manufacturer	4.1.1.2
	Request for Records FDA 482c	Exhibit 5-10
	Sample Collection	Exhibit 5-4
Notice of Inspection		4.2.4
		4.2.4.1
Notice of Sampling		6.2.4.4
Notification of FBI and Us Attorney		5.2.5.4.4
Notifying Receiving Laboratories		4.5.5.5
Nutrition and Nutrition Labeling		6.4.3.4
Nutritional and Allergen Labeling		5.4.6.6.3

O

Objectionable Conditions & Management's Response		5.10.4.3.13
Observations		5.2.3.1.4
Obtaining A Voluntary Embargo		4.2.9.2
OCI Procedures		8.9.1
OCM / EOC Responsibility		8.8.1.1
Office of Criminal Investigation		8.9
		1.9.2.4
Office of Enforcement		1.9.2.3
Office of Regional Operations		1.9.2.2
Division of Federal-State Relations (DFSR)		1.9.2.2.3
Division of Field Investigations (HFC-130)		1.9.2.2.1
Division of Field Science (DFS) (HFC-140)		1.9.2.2.2
Division of Import Operations Policy (DIOP)		1.9.2.2.4
Prior Notice Center (PNC)		1.9.2.2.4.1
Office of Regulatory Affairs		1.9
ORA Map		Appendix E
Office of Resource Management		1.9.2.1
Official credentials, badge		1.6.3
Official Sample		
Private Individual		4.2.6.1

21 CFR 2.10	4.1.4
Official Seals	4.5.4
	Exhibit 4-17
Opening Sterile Sampling Containers	4.3.6.1.3
Organization, FDA	
ACRA	1.9.1
FDA principles	1.8.1
Office of Regulatory Affairs	1.9
ORA field organization	1.9.3
ORA headquarters organization	1.9.2
Organization overview	1.8
Organoleptic Examination	4.3.9
Original Cr & Records To	4.4.10.3.35
Other Acts	2.2.3
Other Government Inspections	3.1.3
	5.4.9
Other Inspectional Issues	5.5.5
Outbreak Determination	8.3.4.1
Outbreaks Associated with Salmonella Enteritidis	
in Eggs	8.3.1.4
Outbreaks Involving Interstate Conveyances	8.3.1.2
Outbreaks on Foreign Flag Vessels	8.3.1.1
Overriding NSD	4.4.10.4.2

P

PAC	4.4.10.3.2
PAF	4.4.10.3.2
	4.4.10.4
	4.4.10.4.3
Packaging and Labeling	5.4.6.6
Packers and Shippers	5.8.4
Parcel Post	4.5.5.7
Parcel Service Shipment	4.4.7.3
Partially Labeled Lot	4.4.9.3
Pathogen Growth Factors	8.3.4.7
Pathological Examination	4.5.3.3
Patient And/ or Consumer Identification on	
Records	5.3.8.6
Payment	
Cost	4.2.8.2

Costs of Supervision of Relabeling other Action6.2.7.9
Labor Cost   4.2.8.4
Method   4.2.8.3
Samples under Court Order   4.2.8.1
                                             4.2.8
Shipping Charges   4.5.6
Payment for Samples   6.2.4.5
Payment Method   4.4.10.3.36
Per Diem and Subsistence
   Documentation   1.2.4
   Foreign Travel   1.2.4
   Late Charge   1.2.4
   Lodging tax   1.2.4
Per Diem Rates
   Commencement   1.2.4.1
   Eligibility   1.2.4.1
Perishable Goods   4.2.9.1
Perishable Products   8.5.7.6
Permit Number   4.4.10.3.37
Personal Safety   5.2.1.2
Personal Safety Alert   5.2.1.3
Personnel   5.4.2
Pesticide Contamination   2.8.2.3
Pesticide Inspection   5.8
   Acreage   5.8.3
   Application   5.8.3.1
   Applicator   5.8.6
   Approach   5.8.1
   Cooperative Activities   5.8.2
   Drift   5.8.3.2
   Growers   5.8.3
   Growing Dates   5.8.3
   Misuse   5.8.3.2
   Packer   5.8.4
   Sampling   5.8.7
   Shipper   5.8.4
   Soil Contamination   5.8.3.2
   Supplier   5.8.5
Pesticide Sample   4.4.10.1.8
   Sample Schedule Chart   Smpl Schdl 3
Pesticides   5.4.7.1.3

Pesticides, Industrial Chemicals, Aflatoxins, &
   Toxic Elements 6.4.3.2
Pharmaceuticals and Medical Devices 3.4.2.2
Photo Identification and Submission 5.3.4.2
Photograph Requests 5.3.4.5
Photographs 4.5.2.4
Photographs 5.3.4
PHS Recommendations Basic Sanitary Practices 5.1.4.2
Physical Hazards 1.5.3.3
Physical Resistance/Threats/Assaults 5.2.1.2.2
Plant Construction, Design and Maintenance 5.4.3.1
Plant Services 5.4.3.3
Plants and Grounds 5.4.3
Poison Prevention Packaging Act 2.2.3.6
Policy (CR 4.4.3
   Fed/State Cooperation 3.1.1
   Consent Decree 2.4.1
   Default Decree 2.5.1
   Compliance Achievement 2.6.1
Port (Point) of Entry 6.7.30
Ports Covered by FDA 6.2.4.1
Ports Not Covered by FDA 6.2.4.2
Possible Contamination Source 8.3.4.6
Post Award (GQA 4.4.10.2.10
Post-inspection Notification Letter 5.3.10
Post Seizure & Reconditioning Samples 4.2.8.1
Post Seizure (P.S.) Sample 4.1.4.7
Post Seizure (Ps 4.4.10.2.11
Postal Box Information 3.2.15.2
Postal Mail Cover 8.9.1.4
Poultry Products Inspection Act 2.7.1.2.3
3.2.1.3
Pre Announcements 5.2.1.1
Pre Inspection Activities 5.2.1
5.2.10.1
5.6.2.1
Precautions 4.5.5.8.7
Precautions during inspections
   Aseptic Technique 5.1.4
   Microbiological Contamination 5.1.4
   Safety Equipment 5.1.4

Sterility	5.1.4
Precautions for Non Clinical Laboratory Inspections	1.5.5.2.3
Preliminary Investigation	8.5.4
Preliminary or Permanent Injunction	2.2.8.5
Premarket Approval	2.9.2.4
Premarket Notification Section 510(K)	2.9.2.3
Premises Used for Living Quarters	5.1.1.9
Preparation for EI	
Complaint	5.2.1
Compliance Program	5.2.1
Guidance Documents	5.2.1
Guides	5.2.1
Postponement	5.2.1.1.3
Pre-Announcement	5.2.1.1
Pre-Inspectional Activities	5.2.1
Recall follow-up	5.2.1.1.2
Preparation and References	5.4.1.1
Preparation and References	5.5.1.1
Preparation of Collection Report	4.4.10.3
Preparation of Detention Notice	2.7.2.3.1
Preparation of FDA 484	4.2.5.5
Preparation of Form FDA 483	5.2.3.1
Preparation of Page 1 (FDA 2289)	2.7.2.3.2
Preparation of Page 2 - 5 (FDA 2289)	2.7.2.3.3
Preparing & Maintaining Digital Photographs	
Evidence	5.3.4.3
For Insertion into Turbo EIR	5.3.4.4
Prescription Drugs	4.2.5.4
Preservation Liquids	4.5.3.1.4
Prints	5.3.4.2.1
Prior Notice Center (PNC)	1.9.2.2.4.1
	6.2.3.5
Prior Notice of Importation of Food and Animal Feed	6.2.3.5
Prior Notice Process	6.2.3.5.6
Prior Notice Reception	6.2.3.5.1
Prior Notice Submission	6.2.3.5.4
Private Individuals	4.2.6.1
Privately-Owned Conveyance	4.4.7.4
Privately Owned Vehicle (POV)	

Official Business   1.2.3
Reimbursement for mileage   1.2.3
Procedure after Hearing "Notice of Release" 6.2.7.7
Procedure after Hearing "Refusal of Admission"   6.2.7.8
Procedure when Conditions of Authorization
    Have Been Fulfilled   6.2.7.5
    Have Not Been Fulfilled   6.2.7.6
Procedure when No Violation Is Found   6.2.6
Procedure when Products Can't be Sampled/
    Examined   6.2.5
Procedure When Violation Is Found   6.2.7
Procedures for Fumigation   4.5.3.1.2
Procedures When Threatened or Assaulted 5.2.5.4.3
Processing Equipment   5.4.7.2.1
Problem Area Flag (PAF)   4.4.10.3.2
Product Code   4.4.10.3.38
Product Description   4.4.10.3.39
Product Disposition   8.5.7
Product/Establishment Surveillance Report Exhibit 8-13
Product Label   4.4.10.3.40
Product Name   4.4.10.3.41
Products Excluded From Prior Notice   6.2.3.5.3
Products Imported under Section 801(D)(3) of the
    FD&C Act   5.1.1.14
  Inspectional Preparation   5.1.1.14.2
  Requirements for BT Act   5.1.1.14.1
Products Requiring Prior Notice   6.2.3.5.2
Products Susceptible to Contamination with
    Pathogenic Microorganisms   4.3.7.7
Professional Personal Contacts   1.6.5.1.6
Professional Reporting System for Vaccine
    Adverse Reactions   8.4.4.1
Professional Stature   1.6.5.1
Profile COMSTAT   Exhibit 5-13
Program Provisions   1.5.1.4.1
Promotion and Advertising   5.4.8.1
Promotion and Advertising   5.5.3
Prosecution
    District Follow-Up   2.2.7.4
    Felony   2.2.7
    Grand Jury Proceeding   2.2.7.3

Information	2.2.7.2
Misdemeanor	2.2.7
Section 305 Notice	2.2.7.1
Protect the Identity of the Source	5.2.9.2
Protecting the Official Seal	4.5.4.4
Protection of Privileged Information	5.2.7.1
Protective and Preventive Measures	1.5.5.2.1
Protective Clothing	1.5.1.3
Protective Equipment	1.5.1
Public Health Service Act (PHS)	2.2.3.7
Public Relations, Ethics & Conduct	1.6
Publications	1.10.2.7

Q, R

Qualifications for Credentials	1.6.3.2
Quality Audit	5.6.2.2
Quality Control	5.4.6.5
Quantity Collected	6.5.5.4
Quantity of Contents	5.4.6.6.1
Radiation Control for Health and Safety Act	5.1.1.10
Radiation Reporting	2.9.2.8
Radioactive Product Sampling	1.5.3.5
Railcars	1.5.3.3.1
Random Sampling	4.3.7.2
Raw Materials	5.4.4
Recall Activities	
Definition	
Depth of Recall	7.1.1.5
Human Tissue for Transplantation	7.2.5
Level of Audit Check	7.3.2.2
Medical Device Notification	7.1.1.7
Medical Device Safety Alert	7.1.1.9
Notification	7.1.1.7
Notification Order	7.1.1.8
Recall Audit Check	7.3.2.1
Recall Classification	7.1.1.2
Recall Completed	7.3.3.1
Recall Number	7.1.1.6
Recall Terminated	7.3.3.1
Recall Type	7.1.1.3

Sub-Account Check	7.3.2.3
Inspection	
Market withdrawal	7.2
Procedure	7.2.1
Recall Decision Follow-Up	7.2.1.1
Recall	
Alert	7.2.7
Close-out Inspection	7.3.3.2
Conducting Audit Checks	7.3.2.4
Food Products	7.2.2
Ineffective Recall	7.3.2.6
Interstate Milk Shippers	7.2.2.1
Medical Device	7.2.3
Monitoring	7.3.1
Procedure	7.2.3.1
Recall Number	7.2.8
Recall Recommendation	7.2.8
Recalls of Human Drug Products	7.2.4.1
	Exhibit 7-1
Recommending Official	7.2.8.10
Reporting Audit Check	7.3.2.5
Sampling	7.2.6
Special Situations	7.4.1
Veterinary Drug Products	7.2.4.2
Recall Audit Check Report	Exhibit 7-2
Recall Number	4.4.10.3.43
Recall Procedure	5.4.8.2
Recall Procedures	5.10.4.3.12
Recall Strategy	7.1.1.4
Recall Terminated / Recall Completed	7.3.3
Recalling Firm/Manufacturer	7.2.8.3
Recalls	7.1
	4.3.5.2
Recalls of Human Drug Products	7.2.4.1
Recalls of Veterinary Drug Products	7.2.4.2
Reconciliation Examination Guidance Part A	5.4.1.4.3
Reconciliation Examination Guidance Part B	5.4.1.4.4
Reconciliation Examinations	5.4.1.4.2
Reconditioned	4.4.10.1.9
	2.3.1.1
	2.6.3

	8.5.7.3
Reconditioning and Destruction	2.3
Reconditioning Devices	8.5.7.10
Reconditioning for Compliance	2.2.6.7
Reconditioning Hermetically Sealed Cans	8.5.7.9
Reconditioning Plastic, Paper, Cardboard, Cloth	
& Similar Containers	8.5.7.7
Reconditioning Screw Top, Crimped Cap,	
& Similar Containers	8.5.7.8
Record Requests	8.8.6
Record Review	
Electronic Filing	6.3.1
Entry Review	6.3.1
Regulatory Authority	6.1.1
Receipt	5.1.1.5
Receipt for Sample	4.1.1.3
	4.2.5
	4.2.5.2
	Exhibit 4-5
Receipt in I.S	4.4.6.2.3
Receipt Issued	4.4.10.3.44
Receipt Type	4.4.10.3.45
Receipts	5.1.1.5
Record Time Screen	6.5.5.9
Recording Complaints/Follow Ups	8.2.8
Recordings	5.3.5
Records	4.5.2.5
Records	5.6.2.3
Records Access under BT Authority	5.4.1.3
Records Accompanying Literature and Exhibits	4.5.2.5
Records Obtained	5.3.8
Redelivery Bond (AKA Entry Bond)	6.7.31
Refrigerated Item	4.5.3.6
Refusal after Serving Warrant	5.2.5.3
Refusal of Entry	5.2.5.1
Refusal to Permit Access to or Copying	
of Records	5.2.5.2
Refusal to Permit Access to Records in Possession	
of Common Carriers	4.4.7.2.1
Refusal to Permit Sampling	4.2.3
Refusal to Sign the Affidavit	4.4.8.2

Refusals	5.10.4.3.14
	4.2.4.2
Refusals of Requested Information	5.2.7.2
Registration and Listing	2.9.1.1
	2.9.3.1
	2.9.4.1
Registration, Listing and Licensing	5.7.3
Approval of Biological Devices	5.7.3.4
Biologic License	5.7.3.3
HCT/Ps	5.7.3.1.1
Laboratories	5.7.3.1.2
Military Blood Banks	5.7.3.1.3
MOUs	5.7.3.2
Registration and Listing	5.7.3.1
Regulations	3.1.2
Regulations	1.10.1
Regulations, Guidelines, Recommendations	5.7.2.6
Regulatory	4.4.10.2.12
Regulatory	
702(a	2.2.1
Decharacterization	2.8.3
Definition	2.2.5
Citation	2.2.5.2
Civil Number	2.2.5.1
Complaint for Forfeiture	2.2.5.5
Criminal Number	2.2.5.3
Denaturing	2.3.1.3
Destruction	2.3.1.2
Device	2.7.1.3.1
Egg Products	2.7.1.3.5
FDC and INJ Numbers	2.2.5.4
Home District	2.2.5.6
Meat Products	2.7.1.3.4
Nolle Prosequi	2.2.5.7
Nolo Contendere	2.2.5.8
Poultry Products	2.7.1.3.5
Reconstruction	2.3.1.1
Seizing District	2.2.5.9
Subpoena Duces Tecum	2.2.5.10
Supervising District	2.2.5.11
Disasters	2.3.2

Regulatory Filing
  Abbreviated New Animal Drug Application
  (ANADA)   2.9.4.3
  Abbreviated New Drug Application (ANDA)   2.9.1.4
  Acidified Foods   2.9.5.1
  Biologic License   2.9.3.2
  Blood Bank Registration and Listing   2.9.3.1
  Classification of Devices   2.9.2.5
  Color Certification Program   2.9.5.4
  Device Registration and Listing   2.9.2.1
  Drug Registration and Listing   2.9.1.1
  FCE Process Filing of LACF/AF Processors   2.9.5.2
  Food Canning Establishment (FCE)
    Registration   2.9.5.1
  Infant Formula   2.9.5.5
  Interstate Certified Shellfish Shippers   2.9.5.6
  Interstate Milk Shippers   2.9.5.7
  Investigational Device Exemption (IDE)
    Regulation   2.9.2.2
  Investigational New Drug Application (IND 2.9.1.2
  LACF   2.9.5.1
    Low Acid Canned Food   2.9.5.1
  Medical Device Reporting   2.9.2.7
    Medicated Feed Mill License (FML)   2.9.4.2
    New Animal Drug Application (NADA)   2.9.4.4
  New Drug Application (NDA)   2.9.1.3
  Premarket Approval   2.9.2.4
  Premarket Notification - Section 510(k)   2.9.2.3
  Radiation Reporting   2.9.2.8
  Requests for GMP Exemption and Variances   2.9.2.6
  Veterinary Medicine Registration and Listing   2.9.4.1
  Voluntary Filing of Cosmetic Product Ingredient   2.9.5.3
    Composition Statement Voluntary Registration
    of Cosmetic Product Establishment   2.9.5.3
Regulatory Notes   2.1
  Electronic Notes   2.1.2
  Format for regulatory notes   2.1.4
  Regulatory entries   2.1.3
  Regulatory note characteristics   2.1.2
  Retention of regulatory notes   2.1.5
  Uses of regulatory notes   2.1.1

Regulatory References and the General Public 1.10.2.8
  Law 1.10.1
  Manuals 1.10.2.6
Regulatory Submissions 2.9
Relabeling 2.4.2
Relabeling 8.5.7.4
Related Samples 4.4.10.3.46
Release of Goods 2.4.7
Release of Information 8.8.4
Removal of Detention Tags 2.7.2.5.1
Repairs 1.6.2.1.1
Report of Analysis 4.1.1.4
  704(d) Sample 4.4.10.3.64
Reportable Observations 5.2.3.2
  Adulteration Observations 5.2.3.2.1
  Other Observations 5.2.3.2.2
Reporting Contacts 8.8.1
Reporting Criteria 5.10.2.1.1
Reporting Sample Collections 4.4.10
Reports 1.10.2.6
Reports of Criminal Activity 8.9.1.1
Reports of Observations 5.2.3
Representatives Invited by Firm to View
  Inspection 5.1.4.3
Request for Authorization to Relabel/Perform
  Other Acts 6.2.7.3
Request for Notice of Inspection 4.2.4.4
Request for Sample Collection 5.3.9
Requesting/Working with Computerized
  Complaint & Failure Data 5.3.8.4
Requesting Computerized Data 5.3.8.4.2
Requests by the Public, Including Trade 1.4.2
Requests for GMP Exemption and Variances 2.9.2.6
Requests for Records Under Section 703 of the
  FD&C Act 5.1.1.7.2
Resealing Conveyances 4.3.4.3
Research Assignments 8.7.1
Research Project Identification Code 8.7.3
Research Project Progress Reports 8.7.4
Reserve, 702(b) 4.1.2
  FACTS Documentation 4.4.10.3.63

Imports	6.5.1
Portion	4.3.3.3
Requirement	4.3.5.1
Resources for FDA Regulated Businesses	5.2.2
Respiratory Protection	1.5.1.4
Response to "Notice of Detention & Hearing"	6.2.7.2
Responsible Firm Type	4.4.10.3.47
Responsibility & Coordination	8.5.2
Responsible Individuals	5.3.6
	5.7.4
Restoring Lot(s) Sampled	4.3.2.1
Retail Stores	8.8.5.5
Retention of Regulatory Notes	2.1.5
Reverse of Tag	2.7.2.4.3
Review of Records	6.3
Reworking	2.4.3
Riots	8.5.5.5
Rodent Contamination	4.3.7.4.2
Collecting Exhibits and/or Subsamples	4.3.7.4.2.2
Examination/Documentation of Contamination	4.3.7.4.2.1
Summary of Sample for Evidence	4.3.7.4.2.3
Rodent or Bird Contaminated Foods	2.8.2.1
Rodents	5.4.7.1.2
Routes of Contamination	5.4.7.1
Routine Biosecurity Procedures for Visits to Facilities Housing/Transporting Domestic or Wild Animals	5.2.10
Routine Requests for Information	3.2.4.3.3
Routing of collection Report	4.4.10.5
Routing of FDA 484	4.2.5.6
Routing of Samples	4.5.5.2

S

Safety	4.3.5
Safety	1.5
Automobile	1.5.2
Animal Origin Products	1.5.5.1
Asphyxiation Hazards	1.5.3.4
Bacteriological Problems	1.5.5.3

Carbadox Sampling	1.5.3.7
Chemical Hazards	1.5.3.6
Electrical Hazards	1.5.3.2
Eye Protection	1.5.1.1
Factory Inspection	1.5.4.2
Hantavirus Associated Diseases	1.5.5.4
Hearing Protection	1.5.1.2
Injury Reports	1.5.7
Inspections	1.5.4
Man Lifts and Ladders	1.5.4.1
Microbiological Hazards	1.5.5
Physical Hazards	1.5.3.3
Protective Clothing	1.5.1.3
Protective Equipment	1.5.1
Radioactive Product Sampling	1.5.3.5
Respiratory Protection	1.5.1.4
Ethylene Oxide	1.5.1.4.2
Fumigation	1.5.1.4.2
Ozone	1.5.1.4.2
Respirator	1.5.1.4.1
Respiratory Protection Program	1.5.1.4.1
Medical Evaluation	1.5.1.4.1
Personal Safety	1.5
Sample Fumigation and Preservation	1.5.3.1
Sampling	1.5.3
Viral and Other Biological Products	1.5.5.2
Safety Precautions	5.2.5.4.2
Sales Records	4.4.7.1
Salmonella Sample Chart	Smpl Schdl 1
Sample	
Reserve, 702(b), Labeling, Documentary Evidence, Witness	4.1.2
702(b) Portion	4.3.3.3
702(b) Portion Collected	4.4.10.3.63
702(b) Requirement	4.3.5.1
704(d) Sample	4.4.10.3.64
Abnormal Containers	4.3.7.5
Accompanying Literature	4.5.2.5
Aseptic Sample	
Controls	4.3.6.5
Dried Powders	4.3.6.2

Handling	4.3.6.4
Procedures	4.3.6.1
Water Samples	4.3.6.3
Water Samples	4.3.6
Authority	
Examination	4.1.1.1
Investigation	4.1.1.1
Notice of Inspection	4.1.1.1
Bill of Lading	4.4.7.2.3
Borrowed Samples	4.5.2.2
Bulk Container Labeling	4.4.9.2
Complaint	4.4.6.3
Complaints	4.3.5.1
Contamination with Pathogenic	
Microorganisms	4.3.7.7
Definition	
301(k) Sample	4.1.4.4
Additional Sample	4.1.4.10
Audit/Certification Sample	4.1.7.1
Dealer	4.2.1
Documentary Sample	4.1.4.2
Domestic Import Sample	4.1.4.8
Food standard Sample	4.1.5
GWQAP Sample	4.1.6
Import Sample	4.1.4.9
Induced Sample	4.1.4.5
In-Transit Sample	4.1.4.3
Investigational Sample	4.1.7
Mail Entry Sample	4.1.7.2
Non-Regulatory Sample	4.1.7.3
Official Sample	4.1.4.1
Post Seizure (P.S.) Sample	4.1.4.7
Undercover Buy	4.1.4.6
Disasters	4.3.5.3
Documentation	4.4.2
Documentation Authority	4.4.1
Documentation of Evidence	4.4.6
Documentation of Interstate Shipment	4.4.7
Documentation Policy	4.4.3
Documentation Procedure	4.4.4
Dry Ice	4.5.3.5

Sample Accountability	4.1.3
Sample Basis	4.4.10.3.48
Sample Class	4.4.10.3.49
Sample Collection	7.2.6
	8.2.7
	8.3.3.1
	8.4.7
Sample Collection During Inspection	5.6.1.2
Sample Collection Reports	6.5.5
Sample Collections	5.8.7
Sample Cost	4.4.10.3.50
Sample Criteria	4.3.7.4
Sample Delivered Date	4.4.10.3.51
Sample Delivered To	4.4.10.3.52
Sample Description	4.4.10.3.53
Sample Flags	4.4.10.3.54
Sample Fumigation and Preservation	1.5.3.1
Sample Handling	4.3.6.4
	4.5.3
	8.3.3.3
Sample Number	4.4.10.3.55
Sample Origin	4.4.10.3.56
Sample Package Identification	4.5.5.1
Sample Records Identification	4.4.5
Sample Schedule	4.3.3
Sample Sent To	4.4.10.3.57
National Sample Distributor (NSD)	4.4.10.4
Sample Shipment	4.5.5
Sample Shipment to Outside Agencies	4.5.5.4
Sample Size	4.3.3
Sample Size	8.3.3.2
Sample Type	4.4.10.3.58
Sampled In Transit	4.4.10.1.10
Samples	6.4.4.3
Samples Collected	5.10.4.3.17
Samples for Pathological Examination	4.5.3.3
Samples for Viral Analysis	4.3.7.8
Samples to Administration Laboratories	4.5.5.3
Sampling	
Preparation, Handling, Shipping	4.5
National Sample Distributor (NSD)	4.4.10.4

Receipt	5.2.4
	5.2.4.1
	5.2.4.2
	3.2.5.2.10
	2.7.3
	1.5.3
	8.8.5.3
Sampling District	4.4.10.3.59
Sampling Dried Powders	4.3.6.2
Bag and Poly-Liner Stitched Together Across	
Top Seam	4.3.6.2.1
Bag Stitched Across Top and Poly-Liner Twist-	
Closed and Sealed with "Twist" Device - Wire,	
Plastic, Etc	4.3.6.2.2
Bags with Filling Spouts	4.3.6.2.3
Sampling from Government Agencies	4.2.7
Sampling (Imports)	
702(b) Reserve	6.5.1
Additional Sample	6.5.1
Collection Report	6.5.5
FDA Coverage	6.2.4.1
FDA Coverage	6.2.4.2
May Proceed Notice	6.2.4.3
Notice of FDA Action	6.2.4.4
Notice of Sampling	6.2.4.1
Official Seal	6.5.1
On-screen Review	6.2.4.3
Payment	6.5.1
Payment for Sample	6.2.4.5
Point of Destination	6.2.4.2
Point of Entry	6.2.4.2
Procedure	6.5.2
Technique	6.5.3
Sampling Lab or Charges	4.2.8.4
Sampling Procedures	8.3.3
Sampling Plan	
Aflatoxin	Smpl Schdl 6
Allergen Samples Schedule	Smpl Schdl 13
Dates and date material	Smpl Schdl 8
Canned and Acidified Food	Smpl Schdl 2
Canned Fruit	Smpl Schdl 7

Color	Smpl Schdl 9
Drug Sampling Schedules	Smpl Schdl 10
Imported White Fish	Smpl Schdl 5
Imports – Coffee	Smpl Schdl 8
Medicated Animal Feed	Smpl Schdl 12
Pesticides	Smpl Schdl 3
Salmonella	Smpl Schdl 1
Veterinary Products	Smpl Schdl 11
Wheat Carload	Smpl Schdl 4
Sanitary Practices	5.1.4.2
Sanitation	5.4.7
Sanitation of Machinery	5.4.5.2
Sanitation Practices	5.4.5.8
Science and Education Administration/USDA	3.2.1.9
Scope of Investigation	3.2.5.2.5
Seafood, Office of	4.5.5.3.3
Seal	
Broken Seal	4.5.4.5
FDA 415a	4.5.4
Metal Seals	4.5.4.6
Method	4.5.4.3
Non-samples	4.5.4.7
Official Seal	4.5.4
Preparation	4.5.4.1
Protection	4.5.4.4
Temporary Seal	4.5.4.5
Application	4.5.4.2
Search Warrant	5.1.1.13.4
Secret Service	3.2.5.2
Section 305 Notice	2.2.7.1
Section 322 of the Public Health Security & Bioterrorism Preparedness and response Act 2002	5.1.1.14
Section 702(e)(5) of the FD&C Act	5.1.1.13
Section 801(d)(3) of the FD & C Act	5.1.1.14
Security	8.8.5.6.3
Segregation	2.4.4
	8.5.7.1
Seizing District	2.2.5.9
Seizure	4.4.6.1
	5.1.1.13
Seizure	2.2.6

Consent Decree	2.4.1
Default Decree	2.5.1
Destruction	2.3
Destruction	2.4.5
Disposition of Rejects	2.4.6
First Amendment Issues	2.3
Reconditioning	2.3
Relabeling	2.4.2
Release of Goods	2.4.7
Reworking	2.4.3
Segregation	2.4.4
Selected Amendments to the FD&C Act	2.2.2
Selective Sampling	4.3.7.3
Bird Contamination	4.3.7.4.4
Chemical	4.3.7.4.5
Criteria	4.3.7.4
Insect Contamination	4.3.7.4.3
Mold	4.3.7.4.6
Rodent Contamination	4.3.7.4.2
Seriously Ill Individuals	4.2.6.2
Sharing Non Public Info with Government Officials	1.4.3
Shipment	4.5.5.8.1
Shipment by Privately Owned Conveyance	4.4.7.4
Shipment of Hazardous or Toxic Items	4.5.5.8.6
Shipper	4.4.10.3.26.6
Shipping	4.5.5
Certified Mail	4.5.5.9
Common Carrier	4.5.5.8
FDA Laboratories	4.5.5.3
First Class Mail	4.5.5.9
Method	4.5.5.6
Notification	4.5.5.5
Overriding NSD	4.4.10.4.2
Outside agencies	4.5.5.4
Package Identification	4.5.5.1
Parcel Post	4.5.5.7
Payment	4.5.6
Routing	4.5.5.2
	4.5.5
Shipping Frozen Item	4.5.3.5.1
Signature	5.10.4.3.21

Signature Policy	5.2.3.1.2
Signing Non-FDA Documents	5.1.2.3
Situational Plan	5.2.1.4
Secret Service	3.2.5.2
Conducting a Special Investigation	3.2.5.2.9
Small Items	4.5.3.4
Small Business Enforcement Fairness Act	1.6.5.1
	5.2.3.1.1
Small Manufacturers	5.6.7
Small Sample Items	4.5.3.4
Sources of Information	1.10.2
Special Information Section	5.4.10.2.3
Special Instructions	6.4.4.4
Special Recall Situations	7.4
Special Regulatory by Product Category	1.10.3
Special Safety Precautions	5.4.1.4.5
Special Sampling Situations	4.3.5
Special Situation Precautions	5.2.10.3
Signing Non FDA Documents	5.1.2.3
Situational Plan	5.2.1.4
Split Sample	4.4.10.1.11
Split Samples	4.5.5.3.2
Standard Narrative Report	5.10.4.3.1
State Operational Authority	3.3
	3.3.1
	3.3.3
State Contacts	3.3.3
State Memoranda of Understanding	3.3.2
State's Operational Authorities	3.3.1
Statutory Authority	5.1.1
Statutory Authority	2.2
702(b)	2.2.1
Amendments to FD&C Act	2.2.2
Codes of Federal Regulations	2.2.4
Device Inspection	2.2.1.3
Drug	2.2.1
Enter & Inspect	2.2.1.1
Examination	2.2.1.4
Food Inspection	2.2.1.2
Investigation	2.2.1
Limitation	2.2.1.4

Other Acts	2.2.3
Record	2.2.1
Sampling	2.2.1
Sterile Devices	5.6.3
Sterilized Equipment	4.3.6.1.1
Storage	5.4.7.3
Storage Requirements	5.2.9.2.2
Storage Requirements	4.4.10.3.62
Stripping (Of Containers)	6.7.32
Sub Account Checks	7.3.2.3
Submitted To	6.5.5.3
Subpoena	1.4.1
Subpoena Duces Tecum	2.2.5.10
Subsamples	4.5.2.1
Substitution	6.7.33
Supervising District	2.2.5.11
Supervision of Reconditioning, Denaturing, or Destruction	2.7.4
Supervisory Charges	6.7.34
Supporting Evidence and Relevance	5.10.4.3.13.1
Surveillance	8.6
FDA 457 Preparation	8.6.2
FDA 457 Routing	8.6.3
Procedures	8.6.1
Survey Sample	4.4.10.1.12
Symptoms Determination	8.3.5.2

T

Tampering	4.3.5.1
Tare Determination	4.3.8.1.1
Taxi 1.2.1.3	
Team Inspections	5.1.2.5
Team Leader Responsibilities	5.1.2.5.2
Team Member Responsibilities	5.1.2.5.1
Technical Assistance	5.1.2.4
	5.6.1.1
Technical Assistance	5.7.2.7
Telephone Communications	
Calling Cards	1.2.8
Calls to Residence	1.2.8

Commercial	1.2.8
Temporary Restraining Order (TRO)	2.2.8.1
Termination of Detention	2.7.2.5
Termination of Research Projects	8.7.5
Testimony	2.2.11
Interviewing Persons under Arrest	2.2.11.2
Miranda Warning	2.2.11.2
Preparation	2.2.11.1
Witness	2.2.11
Testing Laboratories	5.7.5
Tissue Residues	5.9.5
Tornadoes	8.5.5.4
Tort Claims	1.2.2.3.2
Tracebacks of Foods Implicated in Outbreaks	8.3.5.5
Transportation Records for Common Carrier	
Shipments	4.4.7.2
Travel	1.2
Transportation Records	4.4.7.2
Treasury Department	3.2.8
Trial for Injunction	2.2.8.4
Trucks	1.5.3.3.4
Turbo EIR	5.2.3
Type Identification	4.4.10.2
Types of Inspections	5.6.1.3

U

Undercover Buy	4.3.5.5
Unlabeled Lot	4.4.9.3
Use of GFV between Residence & Place of	
Employment	1.2.2.4
U.S. Attorney	3.2.6.1
U.S. Department of Agriculture (USDA)	3.2.1
U.S. Customs and Border Protection	3.2.5.1
U.S. Customs and Border Protection	6.2.3.6.1
U.S. Department of Commerce	3.2.2
U.S. District Court	2.2.6.4
U.S. Nuclear Regulatory Commission	3.2.14
U.S. Patent and Trademark Office	3.2.2.3
U.S. Postal Service	3.2.15
Change of Address Information	3.2.15.1

Postal Box Information	3.2.15.2
Authority	3.2.15.3
Under State Embargo	4.4.10.1.13
Undercover Buy	4.1.4.6
Undercover Buy	4.3.5.5
United States Pharmacopoeia Convention (USP)	3.5.4
Unlabeled or Partially Labeled Lot	4.4.9.3
US Army Corps of Engineers	3.2.3.2
US Army Medical Research & Development Command	3.2.3.3
USDA Acts	3.2.1.3
USDA Complaints	3.2.1.2
USDA-FDA Jurisdiction Chart	Exhibit 3-1
Use of A GFV between Residence and Place of Employment	1.2.2.4
Use of Evidence Gathered in the Course of a Criminal Investigation	5.2.2.6
Use of Evidence Voluntarily Provided to the Agency	5.2.2.7
Use of Tag	2.7.2.4.4
Uses of Regulatory Notes	2.1.1
Utinsels	5.4.5.4
UV Lamps	5.4.5.6

V

Vaccine Adverse Event Report System (VAERS)	Exhibit 8-11
Valid Sample	4.1.2
Vehicles at Receivers	5.4.7.3.2
Vehicles at Shippers	5.4.7.3.3
Veterinary Devices	5.9.6
Veterinary Drug Activities	5.9.2
Veterinary Medicine	5.9
Veterinary Products	8.4.8.2.4
Veterinary Products Complaints/Adverse Reactions	8.4.6
Vet Med Inspection	
Animal Grooming Aid	5.9.7
Authority	5.9.2
BSE	5.9.4

Medicated Feed	5.9.3	
References	5.9.1	
Regulatory Information	5.9.2	
Tissue Residue	5.9.5	
Type A Articles	5.9.3	
Veterinary Device	5.9.6	
Violative Establishments	5.10.4.2	
Violative Inspections	5.4.10.3	
Violative Products	2.6.4.2.1	
Viral and Other Biological Products	1.5.5.2	
Viral Hepatitis & Human Immunodeficiency Virus	1.5.5.2.2	
Viscous Liquids	4.3.8.2.2	
Volume Determination	4.3.8.2	
Volume of Product in Commerce	7.2.8.5	
Voluntary Actions		
Investigator Responsibility	2.6.1	
Compliance Achievement Reporting	2.6.4.2	
DEA Controlled Drugs	2.6.2.1	
Destruction	2.6.2	
Disaster	2.6.1	
Documenting	2.6.4.1	
Reconditioning	2.6.3	
Voluntary Corrective Action	2.6.1	
Voluntary Destruction	2.6.1	
Voluntary Corrections	5.10.4.3.18	
Voluntary Embargo		
Denaturing	4.2.9.2	
Destruction	4.2.9.2	
Inducing	4.2.9.2	
Obtaining	4.2.9.2	
Perishable Goods	4.2.9.1	
State Embargo	4.2.9.2	
	4.2.9	

W, X, Y, Z

Warehouse Entry (WE	6.7.35	
Warrant for Inspection	5.1.1.9	
Warrant Requirement	5.2.2.4	
Waste Disposal	5.4.3.2	
Way Bill	4.4.7.2.5	

Wheat Sample Chart	Smpl Schdl 4
Whitefish, Import sample Chart	Smpl Schdl 5
Wireless Devices	1.5.6
Whole-Bag Screening	4.3.9.1
Working with A Grand Jury	5.2.2.9
Wrecks	8.5.5.6
Written Demand for Records	5.1.1.6
FDA 482a	Exhibit 5-2
Written Notice	5.1.1.3
Written Observations	5.1.1.4
Written Requests for Information	5.1.1.7
LACF / AF Inspections	5.1.1.7.1
Requests for Records Under Section 703 of	
the FD&C Act	5.1.1.7.2
Written Request for Information	5.4.1.2.2
FDA 482b	Exhibit 5-3
Written Request for Records	4.4.7.2.2
X-ray Equipment	5.1.1.10
Field Compliance Testing	2.6.4.2.9
Registration & Listing	2.9.2.1
Youth and Families (ACYF)	3.2.4.2

About the author

Mindy J. Allport-Settle was born in Beckley and raised in Oak Hill, West Virginia. She moved to North Carolina to attend the N.C. School of the Arts for high school and now lives near Raleigh. Following in the footsteps of Gordon Allport, all of her books are built on a foundation of psychology and sociology with a focus on improving some aspect of industry through research and education.

Her career in healthcare began when she was a teenager working as an emergency medical technician. Since then, she has joined the U.S. Navy's advanced hospital corps, worked in organ and human tissue procurement, specialized in ophthalmology, and moved on to serve as a key executive, board member, and consultant for some of the best companies in the pharmaceutical, medical device, and biotechnology industry. She has provided guidance in regulatory compliance, corporate structuring, restructuring and turnarounds, new drug submissions, research and development and product commercialization strategies, and new business development. Her experience and dedication have resulted in international recognition as the developer of the only FDA-recognized and benchmarked quality systems training and development business methodology.

Her education includes a Bachelor's degree from the University of North Carolina, an MBA in Global Management from the University of Phoenix, and completion of the corporate governance course series in audit committees, compensation committees, and board effectiveness at Harvard Business School.

About PharmaLogika

Since 2002, PharmaLogika, Inc has established itself as one of the world's premier consulting firms for Pharmaceutical, Biotech, and Medical Device companies across the globe. In so doing, it has earned the trust of executives in Life Sciences for its integrity, accuracy, and unwavering commitment to independent thought with regard to its products and services as well as those of its customers. Through www.PharmaLogika.com, its involvement in sponsored events, and personal references it has reached millions in print and online. Its mission, to provide flawlessly designed and executed products and services to startups as well as established industry leaders to facilitate their growth from discovery and clinical trial navigation to the commercialization and marketing of their products.

PharmaLogika consults with pharmaceutical, biotech, and medical device quality units to provide third party audits, training, pre approval inspections (PAIs), compliance remediation, and a portfolio of related quality and regulatory affairs products and services. Those products include but are not limited to Quality Assurance Forms, SOP and clinical templates, and the highly successful Integrated Development Training System.

Regulatory action guidance as well as quality systems guidance are delivered as part of its standard products and services. Through the use of highly skilled resources throughout the process, each offering is designed to enact a comprehensive quality systems approach in addressing Quality Assurance (QA) issues. The results insure a close adherence to current Good Manufacturing Practice (cGMP) standards.

PharmaLogika also has a Research and Development OTC line for human consumption that targets alpha 1-antitrypsin deficiency, Fibromyalgia, Restless Legs Syndrome, and Attention Deficit Disorder. A veterinary OTC is currently available that provides canine and feline oral debriding and cleansing agents as well as a stain remover and topical antiseptic. These products combine to provide a strong pipeline of both current and future deliverables.

PharmaLogika

PharmaLogika, Inc.

PO Box 461

Willow Springs, NC 27592

www.PharmaLogika.com

Other books available

Current Good Manufacturing Practices: Pharmaceutical, Biologics, and Medical Device Regulations and Guidance Documents Concise Reference

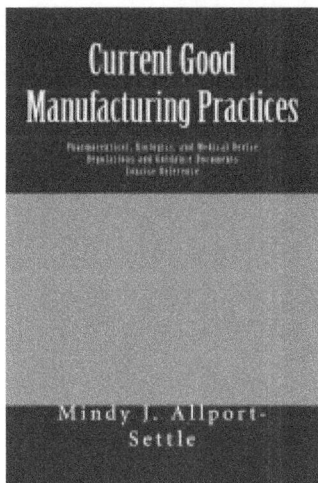

Good Manufacturing Practice (GMP) Guidelines: The Rules Governing Medicinal Products in the Eurpean Union, EudraLex Volume 4 Concise Reference

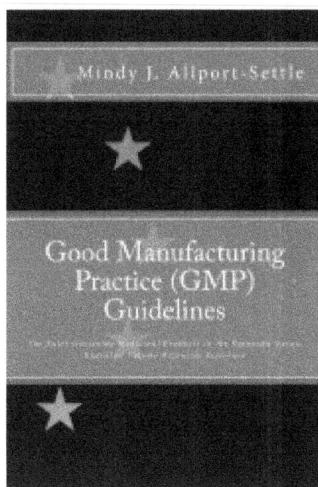

FDA Acronyms, Abbreviations and Terminology: Human
and Veterinary Regualtory Reference

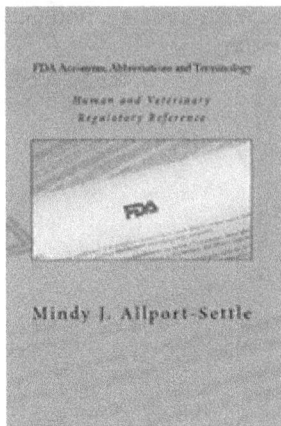

Course Development 101: Developing Training Programs
for Regulated Industries

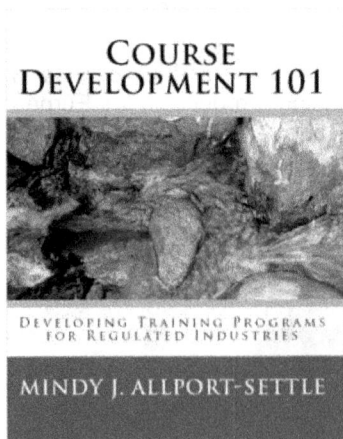

Compliance Remediation for Pharmaceutical Manufacturing: A Project Management Guide for Re-establishing FDA Compliance

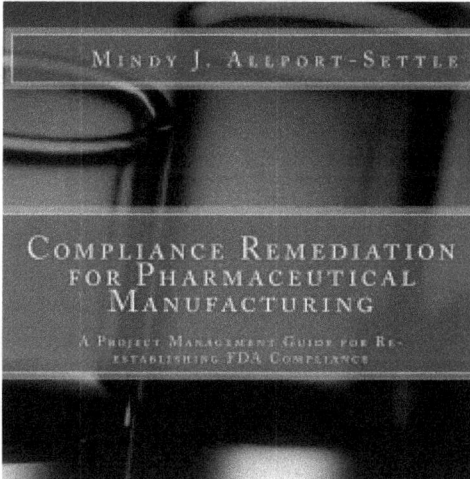

Investigations Operations Manual: FDA Field Inspection and Investigation Policy and Procedure Concise Reference

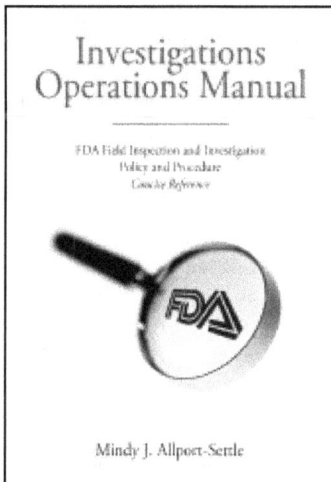

Canadian Good Manufacturing Practices: Pharmaceutical, Biotechnology, and Medical Device Regulations and Guidance Concise Reference

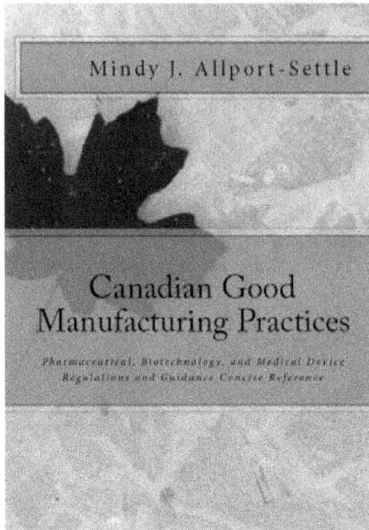

Please visit www.PharmaLogika.com
for additional titles and for a list of resellers
or visit your favorite local or internet book seller.

Companion products and bulk discounts are available at

www.PharmaLogika.com